What People Are Saying About

Consciousness Came First

A truly original piece, drawing across several important dimensions to seamlessly distill down to simple truths that challenge the reader in an engaging, thought provoking and uplifting way. Dr Bilimoria does a wonderful job of providing an intellectually rigorous, yet balanced view on the very nature of contemporary human consciousness. As an academic I find this work to be an absolute delight for the mind and nourishment for the soul. I commend Dr Bilimoria for being a truly original interdisciplinary thinker. The genius of this book lies in its ability to simplify complex philosophical concepts while structuring the text to enable the reader to follow easily. This work can only be described as an award winning magnum opus and a must read for our present times.

Prof Dr Chad Manian, Distinguished Professor at Pebble Hills University, Vice-President of The Academy for Advanced Studies and Chief Academic Officer of Pythia Global Institute – Centre for Peace and Future Policy

I believe we are witnessing a paradigm shift more fundamental and impactful than the Copernican Revolution. Our collective understanding and experience are shifting from consciousness being a mere epiphenomenon of brain function to being the fundamental and ultimate constituent of the universe. As Dr Bilimoria elucidates in his brilliant book, this issue extends far beyond philosophy. Just as placing the Earth at the center of the solar system made planetary orbits nonsensical, this more fundamental misunderstanding has created a metacrisis that will remain intractable as long as we confront problems from a fragmented, 'upside down' perspective. When our perspective is set aright, the unlimited potential of the source and foundation of the universe will infuse individual and collective life with boundless energy, intelligence, creativity, and peace. We'll more easily find solutions to problems, or they will just melt away as our more coherent collective consciousness stops generating them.

Rick Archer, creator and host of the *Buddha at the Gas Pump* interview series

Few authors weave together the threads of science, philosophy, and spirituality as seamlessly as Edi Bilimoria. His work opens a doorway to understanding consciousness as the driving force of life and is an essential guide for those navigating the mysteries of existence.

Bob Cohen, Executive Director, Bhaktivedanta Institute for Higher Studies, Inc, Gainesville, FL

Dr Edi Bilimoria's *Consciousness Came First* is an erudite and penetrating exploration of the nature of consciousness, offering a compelling synthesis of perennial wisdom traditions (as espoused, for example, in theosophy) and modern scientific inquiry. The work provides a visionary blueprint for a new science—one that unites empirical rigor with an expanded metaphysical framework—presenting a holistic worldview capable of addressing the modern spiritual crisis. This crisis, driven in part by the failure of traditional religion and the disenchanted perspective promoted by the scientific establishment, calls for a deeper, more integrated understanding of existence—one that Bilimoria articulates with great clarity and insight.

Dr Pablo Sender, PhD Biological Sciences, author and international lecturer for the Theosophical Society, conducting classes, seminars, and retreats in both Spanish and English

This work is a powerful argument for the scientific and rational view that consciousness is not simply an epiphenomenon of the physical brain, but exists interdependently with the material world. The study of consciousness is the only way to account for various phenomena such as near-death experiences, telepathy, morphic resonance, memories of past lives, and similar observed repeated experiences. The book goes further by showing that our modern discoveries and insights into consciousness are basically reaffirmations of the age-old knowledge about human nature and the cosmos as taught in the ancient mystical and mystery tradition carried over to modern times by later movements such as the Theosophical Society. Culling from a wide range of scientific disciplines (modern physics, neuroscience, medicine, parapsychology, etc.) as well as esoteric literature, this work is one of the most comprehensive attempts to integrate ancient wisdom and modern knowledge.

Vicente Hao Chin, Jr, distinguished theosophist, Founder and Chairman of the Golden Link Schools in the Philippines, former President of the Indo-Pacific Federation of the Theosophical Society, author of *The Process of Self-Transformation*, chief editor of the *Theosophical Digest*, associate editor of the *Theosophical Encyclopaedia*, and compiler-editor of the chronological edition of *The Mahatma Letters to A. P. Sinnett*

This is an astonishingly brilliant and concise summary of our present state of knowledge—and the predicaments it has created. Bilimoria convincingly and definitively refutes the narrow materialism that has prevailed in the sciences. He shows how the Ancient Wisdom, in its many guises, presents answers that science cannot. Bilimoria also reminds us that this ancient knowledge not only provides a vast and comprehensive view of reality but enables us to integrate this vision into a profoundly meaningful human life.

Richard Smoley, author, *The Dice Game of Shiva: How Consciousness Creates the Universe*

Edi Bilimoria is an important figure in Theosophical history. After many years of study within Blavatsky Lodge of the Theosophical Society, founded in London by Madame H. P. Blavatsky herself, he took on significant roles in the team that guides the Scientific and Medical Network which gave him access to the latest advances in consciousness research. The result is a synthesis which no contemporary can equal. HPB who followed scientific discoveries closely would applaud.

Leslie Price, founding editor, *Theosophical History Journal*

My 'Landscape of Consciousness' explores diverse theories of consciousness and their impact on big questions. Understanding consciousness cannot now be limited to selected theories or limited ways of thinking but should seek expansive yet rational diversity. Moreover, issues such as meaning/purpose/value, life after death, AI consciousness, virtual immortality, free will, etc., cannot be understood except in the light of theories of consciousness. This is why I highly recommend Edi Bilimoria's *Consciousness Came First,* a profound, accessible, engaging exploration of perennial philosophy and theosophy. It is not that I endorse these particular theories. It is that I advise understanding them. All who appreciate the landscape of consciousness should have this book.

Robert Lawrence Kuhn, author of *A Landscape of Consciousness,* and creator, executive producer, writer, and presenter of the American public television series, *Closer to Truth*

Edi Bilimoria is a profound thinker, gifted philosopher and a wise man. Consciousness Came First is the essence of his lifelong journey, thinking and research. It is a timely book about timeless philosophy, radical science and eternal truths. It is one of those rare books which explains complex concepts such as

Consciousness, in a readable and accessible way. Read it, you will be inspired!

Satish Kumar, founder, Schumacher College, UK

Consciousness Came First is a wonderful book much needed in our troubled world that has lost it's way down the rabbit hole of materialism, rampant consumerism and greed.

Here we have a clear spotlight on the eternal truths that have been shared since the beginning of time by indigenous peoples who have always lived close to the heartbeat of nature and spirituality. The same truths have been shared by the Ageless Wisdom and the mystical paths of the world's faith traditions; that everything which exists is intrinsically divine and that we are all part of a oneness which permeates all that is.

Furthermore, Dr Bilimoria reminds us that the future of our world is entirely in our hands. With consciousness we are constantly creating our own destinies and that of our world. Karma, or the law of cause and effect, reminds us that the universe continues to evolve and is affected by our thoughts and actions. We are more than our mere physical bodies. As spiritual beings we have a duty to act with compassion, respect and love towards all that is sentient and shares this planet we call home.

Congratulations to Dr Edi Bilimoria on this thoroughly researched and detailed presentation of scientific, spiritual and mystical knowledge.

Pam Evans MBE, founder of the multi award-winning Peace Mala educational project for world peace

CONSCIOUSNESS CAME FIRST

Revealing Evidence for Universal Intelligence and Mind Beyond the Brain

CONSCIOUSNESS CAME FIRST

Revealing Evidence for Universal Intelligence and Mind Beyond the Brain

Dr Edi Bilimoria

The condensed edition of the four-volume *Unfolding Consciousness: Exploring the Living Universe and Intelligent Powers in Nature and Humans* *published by Shepheard-Walwyn in 2022*

London, UK
Washington, DC, USA

First published by iff Books, 2026
iff Books is an imprint of Collective Ink Ltd.,
Unit 11, Shepperton House, 89 Shepperton Road, London, N1 3DF
office@collectiveinkbooks.com
www.collectiveinkbooks.com
www.iff-books.com

For distributor details and how to order please visit the 'Ordering' section on our website.

ISBN: 978 1 80341 973 2 (Paperback)
978 1 80341 988 6 (ebook)
Library of Congress Control Number: 2024948948

A CIP catalogue record for this book is available from the British Library.

Design: Lapiz Digital Services

UK: Printed and bound by CPI Group (UK) Ltd, Croydon, CR0 4YY
Printed in North America by CPI GPS partners

This book is the author's tribute to the illimitable sages of the Orient and the Occident, of all ages, who have epitomized the noblest endeavours of humanity in Philosophy and Science, Religion and Art, thus ever illuminating a Path for mankind through the darkness and turbulence of the mundane world towards his true empyrean abode—should we but listen to their counsel.

Contents

Preface

I regard consciousness as fundamental. We cannot get behind consciousness. I regard matter as derivative from consciousness. Everything that we talk about, everything that we regard as existing, postulates consciousness.

Max Planck[1]

Since the dawn of time, the exhortation by sages of all ages and cultures to *know ourselves* and thereby to know humanity and the world of which we are a part, is, of course, a perennial entreaty. It would be presumptuous, indeed, to profess to have found the ultimate dictum. However, we may claim in this book to have advanced a few small footsteps towards this supreme, but never-ending quest of self-enquiry. But what is meant by the disarmingly simple injunction, 'Know Thyself'? Does it refer merely to our body, which obviously assumes that we are just our body and nothing else? Or does the totality of our being comprise our physical body as well as subtle bodies on higher levels than the physical—giving us form and expression at those levels—all subsumed under the guidance of an overarching master principle? This master principle, we aver, is none other than *consciousness*.

In this book we deal with the unfolding of consciousness—the primary principle of all existence—and its expression through various bodies, subtle and material, on all planes of being from the divine and spiritual to the material and physical. Our scope of enquiry and terms of reference will embrace the universal, ageless wisdom and the corroboration of some of its tenets by modern science.

It is important to understand what is meant by the universal, ageless wisdom. It is known under a variety of terms, all conveying the essential meaning of eternal truth. This corpus of wisdom addresses the oneness of all life and the intrinsic divinity of all that exists. It is the root of all religions, sciences and philosophies that have burgeoned like magnificent fruits from the branches emanating from the central trunk of the Tree of Wisdom. As the term implies, it is the unbounded wisdom of all ages and all times. It can be discerned in the doctrines of the East and the West, from antiquity to the present.

Rudiments of the *eternal wisdom* (another common term) may be found scattered among the folklore of Aboriginal people in every region of the globe. It informs the core teaching of the world's religions and constitutes the matrix of the laws of science as we shall see.

A more specific reference to this universal wisdom is *occult science*. The term *occult* simply means 'hidden' from the Latin 'to hide'; hence the science of the secrets of nature—otherwise known as the Hermetic or Esoteric Sciences. Alchemy is a science to which Sir Isaac Newton (1642–1727) devoted his life. It is an aspect of the universal occult science and means the *chemistry of the invisible or hidden aspect of the laws of nature.*

'Perennial Philosophy' in its original and authentic rendering of *'philosophia perennis'*—which is virtually synonymous with 'Eternal Wisdom'—will be used throughout this book as a generalising term, and for good reason. The humanitarian, philosopher, and physician Albert Schweitzer (1874–1965) likened the evergreen characteristic of perennialism to a tree that seasonally bears the same type of fruit—but never exactly the same fruit, symbolizing that the eternal wisdom—ever One in essence—must periodically be given new expression and communicated in an updated idiom to suit the epoch in question.

The individual expositions of the Perennial Philosophy are extremely varied in their content, focus, doctrines, and expression. Despite the extremely varied language and expression in widely different cultures and historical epochs over vast expanses of time, it is still possible to discern the single thread passing through them that reconciles and unites them as one universal and self-consistent body of wisdom on the common theme of spiritual awakening and realisation.

We have only to look at the unhappy and frustrated condition of large swathes of humanity in the present strife-torn world to realise the disastrous effects of being without any kind of anchoring philosophy or moral compass in life.

Science and technology on their own are poor substitutes—holding out the promise of a sort of gilded dreamland of virtual possibilities which put many people at the mercy of their passions and baser instincts for want of higher principles to guide their lives. The chaotic conditions now prevailing, exacerbated by negative journalism, should convince all about the sheer inadequacy of the predominantly money-oriented, worldly philosophy of life. A higher philosophy must obviously be based on wisdom, not utilitarian ideologies.

This is the reason why none of the sages and philosophers from the Orient or the Occident, from bygone days to modern times, ever allowed the self-realisation of truth to be sullied by the lure of an earthly utopia. They realised that true happiness is never the lot of the spiritually impoverished man. Neither did they waste their time and energy inventing technologies, social schemes or business models for dealing with the turmoil and malice of the world's politics, economic and physical conditions at their own level. With unerring insight, each one of them saw that it was spiritually impoverished man who was himself the prolific spawner of the ubiquitous cruelty and suffering around him; and so it is through man's character, not his acquisitions or possessions, that the way must be found to redeem the world situation: the way of truth, wisdom and love.[2]

People these days tend to attribute the final word of authority to science: what science proves must be true and facts are only facts if they bear the stamp of scientific approval. But it is easy to forget two important points. First, that science is a 'movable feast'. What was once taken as gospel truth gets completely overturned in the light of new discoveries.

The second point is the uncomfortable fact that this very science that has established the groundwork of materialism has also provided the facts that have demolished scientific materialism—namely, quantum science.

The two pillars supporting the edifice of modern science are relativity theory and quantum physics. Both have contributed to shifting the emphasis from matter to mind. Relativity has dispensed with the concept of matter and energy as an intrinsic duality. And it has demolished the notion of matter as the primary stuff of reality since in this new physics it is regarded as crystallised energy.

The discoveries of quantum physics, even more startling than relativity theory to the concepts of classical science, have postulated mind and consciousness as primary qualities (not as the products of matter).

The founder of quantum physics, the German Nobel laureate Max Planck (1858–1947) says: 'There is no matter as such! All matter originates and exists only by virtue of a force which brings the particles of an atom to vibration and holds this most minute solar system of the atom together. We must assume behind this force the existence of a conscious and intelligent mind. This mind is the matrix of all matter.'[3]

Put slightly differently, there is no physical world contained within a non-physical world of consciousness and mind, but *only* the realm of consciousness and mind—the apparently physical world being an appearance, or form, of the mental world.

Have these profound implications percolated into the corpus of the scientific and medical community? Unfortunately, they have not. Despite all these unequivocal assertions about the primacy of consciousness, the vast majority of scientists nowadays still regard a mechanical reality based on matter as the fundamental essence of the universe and existence. The logical corollary of this is that mind and consciousness are the products of material interactions. In the main, establishment science is still steeped in this materialistic and mechanistic paradigm.

Neuroscience and neurobiology are burgeoning subjects of research but this massive effort is based on the materialistic assumption that the physical brain generates consciousness. Are scientists confusing or conflating mind with its mode of expression? In contrast, the knowledge gained through occultism makes our conception much more meaningful and richer by illuminating the inner guiding forces and intelligence which underlie all natural processes and phenomena.

The pursuit of intellectual knowledge for its own sake, which is so fashionably the way of science, is therefore alluring only to those who are not aware of, or choose to ignore, the inner life and the great mysteries of life that challenge us daily on all sides.

The frantic research nowadays to find a physical basis for consciousness is concentrated almost exclusively on the physical brain. That most mysterious and sublimely intimate organ, the heart, is almost completely ignored. The Perennial Philosophy the world over—whether we look to the Greeks, Egyptians or Indians—disseminates the cardinal teaching about the heart in relation to the whole constitution of the human being. How can science, then, ever hope to acquire even a partial, let alone a complete, understanding of consciousness by such neglect of the central organ in the human body?

The central message of this book is that the so-called laws of nature are anything but blind or mechanical in character. They are conscious principles acting as the instrumental functions of Divine Consciousness. The universe is guided by intelligences at different levels of existence. This underlying, all-pervading

Divine Consciousness guides and co-ordinates the exercise of these functions in various spheres. This is why we find perfect co-ordination, harmony, symmetry and *intelligent* direction and control everywhere in nature.

Since time immemorial, sages and seers have all unequivocally affirmed that the material arts and sciences are the mere shadows of Divine Wisdom. One of the most important contributions of occultism is that beside the visible and tangible world which we perceive through our physical senses, there are worlds of a subtler nature. These can also be apprehended by means of subtler faculties which exist in an undeveloped state in the vast majority of people, but which have been refined to an extraordinary degree by those whom we call adepts and occultists.

Another key aim of this book is to show that man and the universe, down to the minutest particle of matter, are imbued with consciousness, goal-orientated evolution and purpose. It argues that once the human state is understood, each human being will realise for himself his unique role in a purposeful universe guided by love and intelligence. It also demonstrates that the human being is a triumph of nature whose future growth and splendour has no limits.

NOTES

1. *The Observer*, 25 January 1931.
2. Marginally rewritten from P. D. Mehta, *The Heart of Religion* (UK: Compton Russell Element, 1976), 125.
3. Max Planck: acceptance speech for 1918 Nobel Prize in physics, quoted in Clifford Pickover, *Archimedes to Hawking: laws of science and the great minds behind them* (New York: Oxford University Press, 2008), 417; Das Wesen der Materie [The Nature of Matter], a 1944 speech in Florence, Italy, Archiv zur Geschichte der Max-Planck-Gesellschaft, Abt. Va, Rep. 11 Planck, Nr. 1797.

Foreword

Consciousness Came First is my personal re-rendering/reinterpretation/elaboration of Dr Bilimoria's trailblazing four volume work *Unfolding Consciousness* published by Shepheard-Walwyn in 2022 and inspired by its intriguing core assertions. It is a book which can, and I hope will, change minds—especially in some circles.

This is not so much a root and branch redaction of a highly erudite, panoramic and original study, but an entire re-focusing of these fascinating ideas aimed at the general rather than academic or specialist reader.

The product of more than two decades research by a modern-day polymath, *Unfolding Consciousness* is arguably one of the most significant works produced in modern times highlighting the body of overwhelming evidence for non-materialist explanations of mind and consciousness. Drawing heavily on the Ageless Wisdom teachings and exploring science, philosophy, religion, and spiritual traditions throughout history, *Consciousness Came First,* covers a vast constellation of ideas and represents a quantum leap in the way consciousness could and should be viewed.

Clearly, re-presenting a monumental work of two-thirds of a million words to around one-eighth of that length presents any author with a number of challenges and difficult decisions. In order to appeal to the non-specialist reader, it has been necessary to omit much technical, historical, and mathematical detail without denting Dr Bilimoria's important core messages. Those wishing deeper insights are encouraged to consult the original volumes.

The central theme of this ground-breaking book is a bold and blunt one: that materialism is a failed worldview. It cannot be countenanced because it is critically flawed in its arguments and assertions. Materialism creates severe boundaries to human perception and therefore consciousness itself. *Consciousness Came First* is effectively a meticulously thought-out manifesto against the physical-only explanation of mind currently espoused by hubristic and myopic science and ultimately embraced unquestioningly by large swathes of the public, goaded, in large measure, by science journalism.

The author makes it abundantly clear that there is no question of castigating science outright. On the contrary, the right use of science—that is, science operating within its legitimate boundaries of investigating physical nature—bequeaths the physical release and material needs of man; it is the worship of science that leads to its wrong use and from there to the attendant ecological, social, and human problems we see all around us. Accordingly, this book bravely challenges the often-unfounded certainties of scientism—that zealous and ubiquitous modern day religious-like cult in which the laboratory and the atom-smasher have replaced the church and the temple.

Scientism is arguably the only example in history of a religion entirely devoid of any spiritual dimension whatsoever—a nuts and bolts, flesh and bone, purely material explanation of everything in the cosmos. Whereas science is a valid attitude, scientism is at best a religion of half-truths. Hence, this book aims to raise science to a higher metaphysic, to win it over to the timeless truths enshrined in the Ageless Wisdom. This is already happening, albeit slowly, through the works of a few brilliant scientists who are expanding the scope of science beyond the confines of classical materialism. For example, the contemporary Nobel physicist Sir Roger Penrose recently stated that the universe has a purpose and is not merely the product of chance.

Consciousness Came First thus represents both a blueprint and radical roadmap for the foundation of a new science—a science of the soul no longer encumbered by the numbing and flawed purely physical explanations of ourselves and the universe. In other words, this will be a re-embracing of occult science. This will be a science unshackled by the restrictions of prevailing paradigms and standard models of physics but liberated into new dimensions of discovery.

Crucially, it proposes and proves the existence of a deeper body of high-grade knowledge in the form of an Ageless Wisdom tradition which extends far back into what is commonly but mistakenly referred to as pre-history. These ancient cosmic principles invisibly permeate most science, religion and philosophy, often unrecognisably because they are frequently disguised and distorted.

These ideas in the form of a Perennial Philosophy have been continually tested and re-tested by the wise down the ages and despite the intervention of temporary fads, have been found to be true.

Consciousness Came First challenges the arrogant modernist assertion that everything we know and understand today is superior to all that preceded it. It clearly demonstrates a common thread of timeless knowledge and universal principles meandering through the centuries and millennia, impacting on traditions along with the rise and fall of civilizations.

This book's title is also its core assertion—that mind creates the material world—not that the material world has first to create the mechanisms required to generate and receive consciousness. For conventional material science mind without a brain is as outrageous as it is unacceptable. But it is one of the kingpins of its older and wiser brother, occult science.

There is another exciting key message embedded throughout this book. This is the assertion—and supporting evidence—that human beings have far more capacity and potential for growth and development than priests or evolutionary biologists will ever admit. Those timeless truths overshadowing every epoch state that human beings are ultimately in charge of their own destinies.

Individual lives are not the result of fate, fortune, luck or chance because these things do not exist. They are weasel words used to describe a process of cause and effect we simply do not comprehend. Nothing happens by accident. Everything has a cause. This is how the universe works. A cosmos without cause and effect is an impossibility.

Consciousness Came First, then, is about evolution via individual effort, persistence and determination. Central to this is a major re-evaluation of precisely what constitutes consciousness, its origins, expression, development, and control—beyond even man's existing and not inconsiderable intellectual attainments.

Before we get overtaken by a tidal wave of hopelessly facile optimism, it is important to caution how Dr Bilimoria repeatedly warns how enlightened minds and original thinkers have always been maligned and marginalised. Throughout every era—and continuing vigorously in contemporary times—creeps a perpetual scepticism of new ideas especially among scientists. Cancellation culture is nothing new. The ancients knew all about it.

Searing cynicism represents a powerful reactionary current and always has done. Rejection, fuelled by mockery, ridicule, and sometimes the loss of livelihood, still remain the cynics' weapons of choice in the second quarter of the twenty-first century. Thankfully, the thumbscrews and funeral pyres now exist only metaphorically.

So, this book is not only a manifesto against condescending modern science; it is a piercing critique of its methodologies and entire mentality. But more than that, it provides us with a means of prising open new vistas, new horizons, and new truths—by using the very thing mainstream science least understands—consciousness.

Tim Wyatt, Loutraki, Gulf of Corinthos, Greece, 17 September 2024

Acknowledgements

I tender grateful thanks to three persons who have been instrumental in bringing this book to fruition.

First and foremost, I am indebted to Tim Wyatt whose journalistic skills along with his deep understanding of esotericism were brought to bear in condensing the four volumes of *Unfolding Consciousness* to a single book aimed at the general reader. Then to Elizabeth Medler for her copyediting and insightful suggestions.

Grateful thanks are extended to my Assistant and PR Manager, Anne Kelly, for communicating with my publishers and managing social media outreach, which have benefitted from her extensive experience as a public speaker and performer on the air in radio and television in addition to her abiding enthusiasm for my book.

Lastly, my thanks are due to my publishers and printers for their care and attention in the production of this book.

Abbreviations and Editorial Method

The Reference Notes appended at the end of each chapter contain abbreviations to save repetition of major works. They are listed below and will be adapted entirely in accordance with the publisher's style and preference.

CW- 'Volume number'	*The Collected Writings* – in Fifteen Volumes by H. P. Blavatsky, compiled by Boris de Zirkoff: Volumes I to VI, The Theosophical Publishing House, Wheaton, Illinois, US, 1988 (Third Edition) to 1975 (Second Edition); Volumes VII to XV, The Theosophical Publishing House, Adyar, Madras, India, 1975 (Second Edition) to 1991.
IU- 'Volume number'	*Isis Unveiled* – Two Volume Set by H. P. Blavatsky, edited by Boris de Zirkoff, The Theosophical Publishing House, First Quest Edition, 1994.
KT	*The Key to Theosophy* by H. P. Blavatsky, Theosophical Publishing House, London, 1968.
NPB- 'Volume number'	*The Notebooks of Paul Brunton* – in Sixteen Volumes by Paul Brunton, Larson Publications, for Paul Brunton Philosophic Foundation, 1984 to 1988.
SD- 'Volume number'	*The Secret Doctrine* – Three Volume Set by H. P. Blavatsky, edited by Boris de Zirkoff, The Theosophical Publishing House, First Quest Edition, 1993.
STA	*The Secret Teachings of All Ages* by Manly P. Hall, Diamond Jubilee

	Edition, Los Angeles, Philosophical Research Society, 1988.
TSGLOSS	Theosophical Glossary by H. P. Blavatsky, Theosophical Publishing Society, 1892.
VS	*The Voice of the Silence* by H. P. Blavatsky, introductory by Boris de Zirkoff, The Theosophical Publishing House, Second Quest Edition, 1992.

For example, *CW*-XII means *The Collected Writings*, Volume XII. Likewise, for *IU*-, *NPB*-, and *SD*-

Note that the initial letter of certain key words is sometimes printed as a capital and sometimes not. The capital letter is used whenever it is necessary to distinguish a word being used in its transcendental sense or as a universal principle from its generic or common use (e.g. Mind contrasted with mind).

Introduction

> *One thing I have learned in a long life: that all our Science, measured against reality, is primitive and childlike—and yet it is the most precious thing we have.*
>
> Albert Einstein[1]

It bears repeating that humanity owes a tremendous debt to modern science and technology. Sadly, this is something quasi-spiritual zealots do not accept and denounce Western science and medicine wholesale while advocating their blanket version of unorthodox medicine and spirituality as the only viable alternative. Such people need to be reminded that it is science that has eradicated much disease through improvements in hygiene, sanitation, and drugs. It is science that has enabled us to communicate with friends and family halfway around the globe.

Thanks to medical science, the dreaded smallpox, which ravaged human populations for millennia and claimed around 300 million lives in the twentieth century alone, was eliminated in the 1970s and is now virtually a forgotten disease.

The scientific method of randomised controlled clinical trials has ensured that complicated surgical techniques, and life-saving drugs administered, can repeatedly be relied upon to perform exactly as required. In our lives, now so reliant upon technology and labour-saving gadgets, science has brought order and dependability, freeing up much time and energy that would otherwise be devoted exclusively to the daily chores needed to maintain our physical existence.

It is also the case that whilst science and technology have solved many apparently unsolvable problems, both in modern times and in centuries gone by, the wisdom which should underpin their application has often been sadly lacking. No sooner is one problem solved than another one is created in its wake. Hence today we see the future of our planet and its inhabitants on a veritable knife edge created by environmental degradation, pollution, and nuclear waste.

Computers, the internet, and mobile phones have contributed enormously to our ability to retrieve vast amounts of information and share it with others anywhere on earth, but what we do with this information is governed by the richness, or otherwise, of our interior lives.

Some drugs (such as penicillin) have achieved spectacular success in combatting sickness. But other kinds of drugs (like Avandia to treat Type 2 diabetes) can produce side effects that outweigh their perceived benefits.

In the mental health field, the efficacy of drug treatment has also become questionable, as in the case of physical ailments. Here we tend to see more dramatic cases of side effects and the drugs (like Selective serotonin reuptake inhibitors) used to cope with depression and similar ailments seem to suppress symptoms rather than addressing the causes of problems. Mental health is now a serious issue.

Let's pose some questions:

Having now jettisoned religion, has mainstream science no ethical signposts? Could it be that it has no higher metaphysical understanding of the implications of its breakthroughs to guide its applications away from areas that are inimical to the human race? And especially when its discoveries fall into ruthless hands?

There is more to life than just the physical world and so by ignoring or discarding wider, deeper and higher dimensions of existence, science is not understanding the whole picture. Could the answer lie not in abandoning the fruits of science in the physical realm but in raising science to a higher standpoint?

If we agree that modern science concerns itself solely with the phenomenal and objective world, then it follows that it must have a limited or partial understanding of the totality of human nature, mind and consciousness.

On the scientific front this book draws upon the latest academic and scholarly books and papers on the paradigm of mainstream science together with its various theories and dicta on the nature of mind and consciousness.

Regarding the Perennial Philosophy, we shall be drawing upon the universal wisdom of all ages, from the West and the East, from antiquity to modern times—but, especially, we shall call upon the good offices of the Theosophical Society which presents a coherent body of knowledge about those portions of the archaic mysteries that are relevant to our age. This knowledge of the unseen realms is the birthright of humanity and has always been scattered far and wide through the folklore, religion, and philosophy of all lands and ages, such as mystic Christianity, Qabbalah and Rosicrucianism in the West, and Zoroastrianism, Buddhism, and the Vedas in the East. But it has been scattered in the form of abstruse doctrines, symbolism and archaic languages to which few have had access.

Theosophy has provided clarity, order, systemic structure, and a rational outlook enabling the student of occultism to acquire a clearer understanding of the processes of manifestation and the laws which underlie the universe, both seen and unseen. But this does not in any way mean that Theosophy has revealed everything. Far from it—the majority of facts have been withheld from an as yet unprepared public. However, that which has lawfully been revealed has provided a good measure of clarity and certainty— but at a price.

The penalty is that many Theosophists regard the classical Theosophical literature as the last word. Thus, progress and research has been vitiated and Theosophy converted by some of its zealots into a new kind of religious dogma which is the last thing on earth that the founders of this great movement intended. To counterbalance this trend towards fossilisation of doctrine is one of the aims of this book.

The contribution of Theosophy and its various offshoots, like the Anthroposophical Society founded by Rudolf Steiner (1861–1925) and the Arcane School of Alice Bailey (1880–1949), to the intellectual and spiritual advancement of the West and the East, and its parental role in spawning related organisations with similar sorts of aims, cannot be overestimated.

That is why we have drawn heavily on the foundational, or classical Theosophical teachings by Helena P. Blavatsky (1831–1891) and her teachers and their subsequent clarification by later generations of Theosophists. These doctrines constitute by far the most comprehensive and thorough expositions on man and the universe that have ever been made available to the public. Previously they had been revealed in secrecy to the deserving few in the sequestered environs of ancient monasteries, temples or Mystery Schools of India, Egypt, the Americas, and Greece.

Amongst later Theosophists, the British social reformer and philanthropist Annie Besant (1847–1933) and I. K. Taimni (1898–1978), professor of chemistry and scholar of yoga and Indian philosophy, deserve particular mention as do the Canadian-born philosopher and occultist Manly P. Hall (1901–1990) and the English writer and philosopher Paul Brunton (1898–1981).

The great British astronomer Sir Bernard Lovell (1913–2012) said, 'A study of history shows that civilisations that abandon the quest for knowledge are doomed to disintegration.'[2] We may surmise that the same would also apply in equal measure to organizations set up for a specific purpose, especially if that purpose be the study and dissemination of spirituality and truth.

This was echoed by the Indian philosopher and international speaker Jiddu Krishnamurti (1895–1986) in 1929: 'I maintain that Truth is a pathless land, and you cannot approach it by any path whatsoever, by any religion, by any sect [...]'

Krishnamurti maintained that no organisation or outer structures can lead man to spirituality because any organisation created for such purpose soon becomes a crutch or a bondage and 'must cripple the individual, and prevent him from growing, from establishing his uniqueness, which lies in the discovery for himself of that absolute, unconditioned Truth.'[3]

If the Theosophical Society were to establish any one teaching (including Blavatsky's) as exclusive and authoritative, the individual search for truth, freedom, and liberty would necessarily be stifled.

Blavatsky's statement was reinforced by her disciple, Annie Besant: 'To proclaim one person as an infallible authority on a subject unknown to the proclaimer, is to show fanaticism rather than reason. The Theosophical Society [...] may be injured by the blind zeal of those who pin their faith to any one investigator and denounce all the rest.'[4]

Likewise, I. K. Taimni warns us that 'the Eternal Wisdom is a transcendent Reality which cannot be poured into a mould, preserved [in formaldehyde] and then worshipped as a fetish.'[5]

Sadly, this is what has sometimes happened. Instead of striving ever forward and seeking new avenues of research and updated, modern terms of expressing the ageless wisdom corroborated by the latest science, the Theosophical Society has rested on its laurels with the result that the vital and living message has become congealed into dead-letter modes of interpretation: the outer form of words has predominated over their inner meaning.

Blavatsky herself told us that *The Secret Doctrine*, her magnum opus, 'lifts only a corner of the veil—that it must be drawn out and lived through the human soul.'[6]

As we would expect, such attitudes are prevalent in spiritual societies that by their very nature attract members of devotional dispositions. But the solution to the problem lies in an organisation that encourages networking amongst its members along with special interest groups, rather than hierarchical structures with an identifiable head, thereby placing no constraint upon individual freedom of enquiry. Such let it be said, is the Theosophical Society when true to its mission. Another fine example is the modern Scientific and Medical Network.[7]

To treat any book, teacher, or doctrine as a creed and absolute standard of authority is tantamount to a betrayal of the Perennial Philosophy which is larger than any one of its manifold expressions.

NOTES

1. *Einstein in Brief*, American Institute of Physics <https://history.aip.org/exhibits/einstein/inbrief.htm> accessed 1 April 2024. Quoted also in Jack Brown, 'Reminiscences – I Visit Professor Einstein', Ojai Valley News, Ojai, California, 28 September 1983, <http://www.blavatskyarchives.com/brown/jackbrownoneinstein.htm> accessed 1 April 2024.
2. *The Observer*, 14 May 1972.
3. Mary Lutyens, *Krishnamurti: The years of awakening* (London: Rider, 1984), 272–5.
4. Annie Besant, 'Investigations into the Super-Physical' (Theosophical Publishing House, December 1913), Adyar Pamphlet No. 36 http://hpb.narod.ru/InvestigationsSuper-physicalAB.htm accessed 25 October 2019.
5. I. K. Taimni, *Man, God and the Universe* (Adyar, Madras: Theosophical Publishing House, 1969), 379.
6. *Insight – Journal of the Foundation for Theosophical Studies* (UK, November–December 1999).
7. The Scientific & Medical Network <https://scientificandmedical.net> accessed 1 April 2024.

Part One

Who or What Am I?

Indissolubly linked with thought and action, love is their common mainspring and, hence, their common bond.

Louis de Broglie[1]

Atheism is so senseless and odious to mankind that it never had many professors.

Sir Isaac Newton[2]

The commonest failing is the sectarian spirit in which people diminish themselves by rejecting others.

Gottfried Wilhelm von Leibniz[3]

By materialists I mean those who are, so to speak, mentally retarded by virtue of wilful refusal, despite all evidence, to admit the possibility of nonphysical life.

E. Lester Smith FRS[4]

I know from personal experience that I am not just the body that I use. After all, I can make my body perform acts that it detests or likes doing. Who, or what, is the entity, the 'I', that is so acting on my body? Similarly, I know that by an act of will, I can change my emotional state, such as managing and redirecting my anger, dispelling depression, or calming my anxiety. Significantly, I also know that, despite the difficulties that I may have in my personal life, I can, as a deliberate act of will, select my thoughts and to some extent control them to create an optimistic outlook, rather than succumb to bleak pessimism. I also know that I may choose to do the wrong thing or speak hurtful words, even though I know in advance about the troubles that will ensue.

Infrequently, I can sense a sort of blurring of my subject-object awareness, in the sense that what I do, and the act of doing, seem to be fused—popularly known as being 'in the flow', or living 'in the present moment'. This brief introspection shows me that I am not the same as my body or my emotions and certainly not my thoughts. My instruments of expression physically, emotionally, and intellectually are at a lower level and therefore not the same as the 'I' who uses them.

Mainstream science has concentrated heavily on the outer, physical aspect of man to the virtual exclusion of his inner informing principles, namely his mind and spiritual nature. A physical mechanism can be investigated by systematically taking it apart and examining its systems, components, and sub-units down to the minutest detail. The whole is not greater than, but merely the sum of its component parts. So, the mechanistic outlook, with its associated methodology of reductionism, has turned out to be the 'machine paradigm' of science regarding the universe and man.

Because this has become so entrenched in establishment science and taken on the role of an incontestable dogma, it would be useful to examine the philosophy and ensuing implications of science, in order to identify the deficiencies in its exclusively materialistic worldview. This reinforces our case for the indispensable need for the psycho-spiritual and esoteric sciences to complete and complement (not replace) the picture of mainstream science.

Science cannot be divorced from its underlying philosophical foundations or from the technology and associated institutions and industries built around it. These are strongly coupled with commercial, financial, and political structures. Some systems and institutions have become so large and bureaucratic that the human element is starved. In blanket terms, it cannot be argued that science is 'making the world a better place' or that science is 'making the human being happier or more human'. In many areas we are immensely grateful to science for the alleviation of material poverty and the health of the physical body. But there remains the overriding question that science, in the main, fails to address: 'Who, or What Am I?'

Despite the phenomenal advances in science one sees a parallel, exponentially worsening, trend of *inner* poverty, one might say 'soul poverty'—a starvation not of the belly but of the soul and spirit as evinced in all kinds of social disorders such as drug taking, alcoholism, sexual violence, and barbaric or lewd entertainment, where the individual attempts to fill an inner vacuum with sensory distractions. As cogently argued in *The Metaphysics of Technology*, there are two phases of technological development and determinism: 'anthropogenic'—dependence without control and 'autogenic'—when technology will become self-making and self-evolving.[5] The danger obviously lies with the latter—advances in technology (in some sectors, like computing, at an exponential rate), going hand-in-hand with a rapidly declining quality of life.

Of course, it is not technology *per se* that is the cause of the current malaise stemming from technological and psychological stress due to information overload, but human beings who have engineered the situation. So, the solution to the problems of excessive technology is not a technical one but lies with the human being in terms of his ability to forge healthy relationships with his fellow beings, his role in society and in the world, his relationship to nature, and the fulfilment of his inner nature—all underpinned by understanding himself.

The Mind-Brain-Thought Problem: Even Nobel Scientists Disagree

Nothing is more basic to our lives as thinking beings and nothing, it seems, is better known to us than our personal and individual experience. But the ever-expanding reach of mainstream science proposes that everything in our world is ultimately physical. Among the most intriguing problems in neuroscience is the challenge of fitting consciousness into the modern scientific worldview and showing that it is nothing but neural activity in the brain.

The problem goes back a long way, in fact, well into the nineteenth century. It was the German scientist, philosopher, zoologist, and politician, Karl Christoph Vogt (1817–1895), who first mooted the idea that the brain generates consciousness. His well-known quote of 1846 was, 'The brain secretes thought as the stomach secretes gastric juice, the liver bile, and the kidneys urine.' Hence, Vogt continues, 'thoughts stand in about the same relationship to the brain as bile to the liver and urine to the kidneys.'[6]

Despite the fact that the thoughts and language of large portions of human society (high ranking politicians included) are fit for the urinal, does the reader truly think that thoughts expressing nobility, sublime feelings, deepest love, universal compassion, scientific discovery and artistic creation from the likes of Buddha and the Christ, or Plato (*circa* 427–348 BC), Leonardo da Vinci (1452–1519), Shakespeare (1564–1616), Newton, and Mozart (1756–1791) can all be compared entirely to the brain equivalent of urine or bile production? Some Nobel laureates seem to be in no doubt.

Arguably, the supreme modern exponents of the mainstream scientific theory about mind and consciousness are the English molecular biologist and neuroscientist Francis Crick (1916–2004), who co-discovered the deoxyribonucleic acid (DNA) spiral with the American molecular biologist and geneticist James Watson (*b.*1928). Crick opens his book, *The Astonishing Hypothesis: The scientific search for the soul* with this manifesto: '"You", your joys and your sorrows, your memories and your ambitions, your sense of personal identity and free will, are in fact no more than the behaviour of a vast assembly of nerve cells and their associated molecules.'[7] Later in the book he continues along the same lines with, 'a person's mental activities are entirely due to the behaviour of nerve cells, glial cells, and the atoms, ions, and molecules that make them up and influence them.'

Few amongst us would deny the role of neurobiology, but do neural mechanisms *of their own accord* supply the final answer to all our joys and sorrows, memories and sense of self?

Other Nobel laureates are not so sure; for example, the American neuroscientist and philosopher of mind, Gerald Edelman (1929–2014), says: 'To reduce a theory of an individual's behaviour to a theory of molecular interactions is simply silly [...] when one considers how many different levels of physical, biological, and social interactions must be put into place before higher order consciousness emerges.'[8]

The problem is the reductionist method of materialist science where the whole must be understood by reducing it to its most elementary parts. When taken to absurd extremes, reductionism ceases to be a valuable technique in science and becomes an ideology upheld by faith rather than reason.

According to the Darwinian theory of evolution, we humans are the product of the interplay of chance mutations of genes and natural selection, which chooses the most favourable genes for our survival. Regarding consciousness, the accepted dictum in mainstream neuroscience is that our brain is the thought producer or generator of our thoughts and that our consciousness is purely the result of brain neuronal activity and nothing else.

How does a combination of genes conspire to produce staggering talent? Natural selection is decidedly a slow process over aeons but, somehow, inexplicable talent appears out of the blue in a family invariably of mediocre background. Is there a gene, or combination of genes that results in something so sudden and so specific as high musical talent?

Let us take this a step further and consider the sudden onset of musical genius. Arguably, the finest example is Wolfgang Amadeus Mozart, a child prodigy caught by his father composing a difficult harpsichord concerto when scarcely aged five. To this day, musicians are awestruck by Mozart's insights into human nature but where did such profound insights come from?

Turning to the world of science, Isaac Newton was born to a farming family 'wholly without distinction and wholly without learning.'[9] Yet this diminutive and prematurely born child, given little chance of survival, whose father was unable to sign his own name and whose mother was only barely educated, became one of the most celebrated scientists and a figure of universal significance for mankind, living to the ripe old age of eighty-five. His work shook science to its foundations.

So why did this particular individual, amongst all others, produce colossal tomes on mathematics, physics, optics, theology, alchemy, philosophy, chronology, economics, geography, plus voluminous correspondences and administrative papers, where he held high office as Warden, then Master, of the Royal Mint, and President of the Royal Society? Why, then, this sudden onrush of supernal genius when aged 22–24, when he made revolutionary discoveries and formulated the universal law of gravitation—all triumphs unequalled in the history of science?

Let us suggest that the answer lies entirely in our genes and neurons and that the essence of life, mind and consciousness is solely material and physical. In that case we are forced to deny the existence of any non-physical creative agency in the universe and therefore we must ascribe all creativity and genius to pure chance to the fortuitous concurrence of genes that, in the case of Newton was somehow 'switched on' in his early twenties.

If this is the case then, it is not unreasonable to ask: what ultimately is a gene? Surely just a physical chemical compound of atoms and molecules operating according to the well- known laws of physics and chemistry. Then, regarding Mozart, how do such complex arrangements of chemical compounds translate what is entirely vibrations in the air of certain frequencies—pure sounds—into sensations that we experience as sublime music that can move us to tears?

Is it the case that the brain and the mind are intimately correlated or are they one and the same thing? If the answer be the latter, it would be no different in principle to asserting that a piano and a pianist are the same thing—or that the wood and strings of a piano secrete sound with no need of a pianist.

There are innumerable books, learned papers, and courses by international scientists and neuroscientists on the subject of brain and mind. Two of the best examples of modern expositions about the mind-brain problem in neuroscience are the joint courses[10] by two American scientists: Steven Novella (*b.*1964) and Michael Shermer (*b.*1954). The ideas of Novella and Shermer, are regarded as entirely typical of the standpoint of the vast majority of scientists and neuroscientists.

Shermer asserts that 'mind is just a word we use to describe neural activity in the brain', a rigid orthodox stance reinforced by Novella that, 'one of the premises of this course is that we are our brains' and who further asserts, 'the left hemisphere has the ability to speak and understand language.'

How scientific is it to ignore the massive body of evidence from academic scientific, peer-reviewed journals that contradict the dogma of brain = mind? Later in their courses Novella and Shermer flatly dismiss not just some but all the evidence from parapsychology and *psi* research.

Is it scientific to premise a course on an assumption that is merely stated without supporting evidence and simply taken for granted? What is being promulgated as established and undisputed scientific fact is the theory and belief that nothing exists in the universe except physical matter. This attitude is not one of science, but an ideology propped up by the 'church of scientism'—the equivalent of religious propaganda. Hence, if matter is taken as the only or fundamental reality, then all life, processes and phenomena can be explained as manifestations or results of matter. From this it is obvious why materialists must deny the existence of spirit, or any kind of non-physical occurrences, and rely upon purely physical explanations for all phenomena, including the so-called paranormal.

By paranormal we mean occurrences and phenomena that are non-physical, being beyond the range and reach of ordinary human capabilities and not amenable to conventional scientific explanation.[11] A similar term is *psi*, the 'anomalous processes of information or energy transfer such as telepathy or other forms of extrasensory perception that are currently unexplained in terms of known physical or biological mechanisms.'[12] Although mainstream scientists remain extremely sceptical about such claims, there is a typically high level of belief in the paranormal among the general population.

Novella claims that 'researchers have not been able to convincingly show that psi phenomena exist', furthermore, that 'psi research [...] has been going round in circles and not progressing at all. It has yet to develop a single repeatable demonstration of psi.' Reproducibility has virtually assumed the status of a religious commandment in science. But is everything precisely reproducible?

With physical experiments, if all parameters (initial conditions, boundary conditions, experimental protocols) are kept exactly the same, we can rightly mandate that the results must be exactly the same. But with *psi* experiments the parameters are vastly greater, subtler, and more sensitive to influence than physical experiments (the emotional state of the experimenter being one).

From the voluminous body of evidence we may cite such diverse lines of research as the investigations of direct mental interaction with living systems;[13] studies on pets that appear to know when their owners are returning home;[14] research into hypnotic telepathy conducted in Russia;[15] and studies on remote viewing.[16] The statistical significance of *psi*, shown by the meta-analyses (the combination of results of multiple scientific studies) of the American parapsychologist Dean Radin (*b*.1952) and collaborators, is impressive.[17, 18] Another area of contention concerns the whole field of survival after death. That consciousness is not extinguished at death is an unequivocal key tenet of the Perennial Philosophy of all ages since time immemorial. There is also considerable evidence to be found in extensive literature on the subject of spiritualism, as in the annals of the British Society for Psychical Research in England, founded in 1882, and the first Society to conduct organised scholarly research into human experiences that challenge contemporary scientific models.[19]

Near-death experiences have a long history and are well documented—increasingly, these days, with rigour and insight from eminent doctors. One example is the American neurosurgeon Dr Eben Alexander MD (*b*.1953). He contracted a severe brain infection that left him in a coma close to death and in a vegetative state with no capacity to create thought since the neocortex (the part of the brain involved in higher functions, such as sensory perceptions, conscious thought, and in humans, language) had effectively shut down. Amazingly, he fully regained consciousness with a coherent and profound set of memories during the period of the coma. In his book *Proof of Heaven*, Alexander states, 'My experience showed me that the death of the body and brain is not the end of consciousness [...]. What happened to me while I was in a coma is the most important story I will ever tell. But it's a tricky story to tell because it is so foreign to ordinary understanding.'[20]

Another instance of the separation of consciousness from the body is that of Dr Rajiv Parti MD (*b*.1957), a world-class cardiac anesthesiologist. He was the last man to believe in 'beyond the brain' states of consciousness—until his near-death experience on an operating table, where he watched his own operation from the ceiling in accurate detail, even recalling a joke told by the doctors during his surgery. His near-death experience led to a profound spiritual awakening that transformed his career, lifestyle, and most fundamental beliefs, as shown in his book *Dying to Wake Up*.[21]

Before his life-changing experience, he was both selfish and materialistic, living in a large mansion, running expensive cars, and adopting a high-pressure lifestyle—not to mention his callous attitude to many of his patients and abusive pressure meted out to his eldest son. Near-death proved to be a dramatic awakening for him. Not only did he lose interest in material things, which resulted in downsizing his home and disposing of his expensive cars, he completely re-orientated his life towards spiritual values. Moreover, he took up meditation to gain access to inner guidance and practised immense kindness and love towards his wife, son, and his friends. There is a wealth of information on the life transforming and soul-affirming nature of numerous NDEs from modern scientific studies.

The Indian sage and philosopher Jiddu Krishnamurti constantly admonished those who argued dogmatically from a particular point of view—whether as a scientist, a religious leader, politician, or in any other role, by reminding them that we are all first and foremost human beings, not the labels and concepts we accumulate about ourselves.

The Paradigm of Science – Its Ideology, Assumptions, and Beliefs

Attempting to explain consciousness on a purely physical (neurological) basis, another powerful driving force must be noted. It is the phenomenal advances in natural science itself over the last century. It would be difficult to identify any other realm of human endeavour that has progressed by such leaps and bounds over this period as science and associated technology. This has unfortunately engendered a hubristic attitude of omniscience in the scientific community at large. Technology is the application of science but because of their close coupling, it is argued on the analogy of hardware and software in computers, that there ought to be an equivalent technological basis to mind.

The contemporary American philosopher Josh Weisberg takes on one of the most intriguing explanatory problems of our time: the challenge of fitting consciousness into our modern scientific worldview, of taking the subjective feel of conscious experience and showing that it is just neural activity in the brain. His book, *Consciousness: Key Problems in Philosophy* concerns the complexity of current debates on consciousness.[22]

He presents the range of contemporary responses to the philosophical problem of consciousness and also systematically deals with philosophical theories of consciousness. The extreme points of view are held by the dualist and other non-reductive theorists, who hold that our basic physical ontology must be enlarged to include the intrinsic features of consciousness. Conversely, identity theorists and functionalist theorists hold that consciousness can be accommodated into the standard (i.e., materialistic) scientific picture provided that the philosophical platform is suitably modified. It would seem, therefore, that modifying the philosophical platform to accommodate the scientific picture is tantamount to moving the goalposts to achieve the intended result.

One of the greatest strengths of science is its unimpeachable adherence to discovering truth and its painstaking rigour in doing so. Within the scientific camp itself, and as a result of its own labours, has emerged a new light: the science of qualities which, in the spirit of Goethean science, puts the emphasis on the *whole system* of the organism rather than the component parts. This new science, along with other developments in the field, like quantum theory and the theory of morphogenesis (considered later), is slowly but inexorably uplifting the corpus of natural science towards a higher metaphysic and a wider perspective.

Since antiquity, philosophers and scientists have realised that all natural processes have an aspect that is rational and quantifiable. But it was Galileo's crowning achievement to delineate a methodology of natural science in quantitative terms, i.e., in terms of a systematic study of number and measure. Ever since then, mainstream science has focussed almost exclusively on the measurable, mathematically expressed—mass, velocity, momentum, etc.—and ignored the domain of experience known as 'qualia', such as beauty of form and texture, since they could not be quantified with certainty or precision. This resolve towards the quantification as a characteristic of our times has resulted in a science well adapted to the machine-like aspects of the world.

In modern times, another weakness of the science of quantities is its failure to provide satisfactory predictions for complex systems. Weather patterns and an organised cloud of migrating starlings are two of many such examples that resist solutions from a science with a strongly deterministic and reductionist methodology. Why? Because, in *every* sense, the whole is greater than the sum of its parts.

Healing the Split

The recent discovery of deterministic chaos in dynamical systems (e.g. weather patterns) allow a reconciliation of the two apparently contradictory aspects of certain categories of natural processes that are intelligible but apparently unpredictable. With this discovery, scientific knowledge, originally seen to make possible the prediction and manipulation of nature, appears now to point towards a new relationship with the natural world based on sensitive observation and participation, rather than control. This requires the cultivation of a new type of science, a science of qualities, albeit with its roots in the past.

Up until the mid-nineteenth century, before becoming mesmerised by technology, the imagination of humanity was still learning from and enamoured by nature.

Towards a Science of Qualities

The consequence of this new way of looking at nature and complex systems is a movement beyond mechanism and reductionism. The action of a machine can be fully described in terms of the forces acting between separate parts, which are coupled to one another in unique and specific ways to achieve a particular function. But when this type of mechanical explanation gets extended or extrapolated to the action of dynamical systems and living processes, such as molecular reactions, the results are unsatisfactory.

This is because of the unwarranted assumption that understanding the properties of the individual reacting molecules is sufficient to provide an explanation of the overall reaction and the new molecules, or biological tissues, that are produced from it.[23]

Entropy and syntropy are opposite concepts that describe the level of organisation or disorder in a system or organism. Entropy is a quantitative measure of the degree of internal disorganization. Conversely, rather than generating disorder via increasing differentiation, syntropy draws individuals and systems together based on their similarities. In a certain sense, syntropy can be regarded as the action of love.[24] It is quite startling to find the word 'love' used in a strictly scientific context but this is indeed the case. Furthermore, it is immensely reassuring to find that even a handful of top scientists are waking up to the fact that human beings are not complex machines programed by a wet computer, otherwise known as the brain, but on the contrary, spiritual beings animated by the exhaustible force of love. The implications of this are inestimable, not just for science, but for our whole culture.

One cannot but notice the twists and turns of materialist philosophers and scientists in their resolute attempts to avoid invoking any notion of non-physical causes or mind and consciousness transcending the brain, which would obviously imply higher states of matter and being—let alone any question of divinity. 'Physicalism' is the fashionable term used to replace materialism. But surely this is all just a game of semantics as both terms denote essentially the same thing: that all that exists, including mind, is ultimately physical and therefore can be traced back and reduced to the most elementary physical building blocks.

The central weaknesses, therefore, in the materialist position are twofold:

1. The unwarranted assumption about the primacy of matter. Re-labelling 'materialism' as 'physicalism' does nothing other than move the 'hot potato' from one hand to the other.
2. The corollary to this is the assumption that the physical senses are the only senses that a human being possesses. The vast corpus of mystical and esoteric literature the world over, since time immemorial, attests to faculties transcending the physical, which therefore enables man to explore higher dimensions of being and states of consciousness.

This closedness of the physical has had a major, not to say devastating, impact on the progress of the scientific investigation of mind for it has resulted in the 'closedness of the mind' by limiting science (including, of course, medicine and psychiatry) just to the physical domain. Scientific research in general nowadays is limited by highly vested interest groups—vested not only in the sense of acquiring funding or research grants, but the phobic attempt to uphold materialism and defend the paradigmatic fortress of scientific materialism against all other complementary viewpoints. This is no exaggeration.

It is a matter of common experience and observation that those scientists who are solely wedded to the materialistic paradigm will protect the notion of physical causal closure at all costs and overtly display a 'closure of mind' in regard to new evidence that seems to violate their fixed position. Such new evidence could be, for example, authenticated cases of non-drug-based cures for cancer, clear evidence of near-death experiences entailing vivid recall of specific events, despite the patient being clinically 'brain dead', and authentic cases of telepathy.

The list could go on but the common feature about the response of the scientists is their fearful and defensive attempts to maintain the status quo. When normal physical facts fail, they will frantically attempt to *explain away* the evidence by any means possible, such as using statistical artefacts, or coming up with all manner of irrelevant counter-arguments or resorting to the usual overriding accusations of poorly designed experiments, experimental error, and fraud.[25]

But when all else fails, the final resort is to overbearing authority. Indeed, several of the 'paradigm police' are immensely powerful authority figures in the world of mainstream science—Sir John Maddox FRS (1925–2009), the British chemist and editor of the prestigious science journal *Nature*—who would think nothing of crushing the career of anyone who dared to oppose his ideological beliefs, if such opposition were to result in new ways of viewing the universe or holistic forms of medical care, for the simple reason that immense sums of money and vested interests are involved. For example, allopathic medicine would be threatened by public perception of non-drug-based forms of treatment like acupuncture or homœopathy.

Strong and powerful egos are behind much of mainstream science and medicine in large multi-national corporations, especially the pharmaceutical industry, where massive finances and prestige are involved, more than the proponents are prepared to admit or are even aware of. *Deadly Medicines* and *Organised Crime* and *Bad Pharma* are authoritative and well-researched books painting a similar picture of systemic corruption in the pharmaceutical industry which has effectively 'bought out' the health system: meaning journals, medical education, doctors and psychiatrists, patient groups, governments, and regulators.[26] A more covert, but no less ruthless, operation of the paradigm police is to be found on Wikipedia. Scientists whose ideas are perceived as posing a threat to mainstream practices, and the funding that goes into them, have had major portions of their web pages slanted or eradicated.

A classic instance is Rupert Sheldrake's talk 'The Science Delusion', at TEDx Whitechapel in 2013, which was removed from circulation by the TED community after being aired in response to protests from two hardcore materialists in the US.[27] Ironically, before the banning the video had a modest 35,000 views. Since then its clones have been watched over five million times and it has also been dubbed in Russian and has subtitles in over twenty languages.[28]

What do we think these scientists should be saying when faced with unknown territory and anomalous evidence? Why not start with humility and the simple words: 'I simply do not know, but how interesting. Therefore, I will listen first and try to find out in the true spirit of scientific enquiry, without any prejudice or preconceived theories.' Alas, such thoughts hardly enter their minds or equivalent words pass through their lips! How can the new be discovered from the standpoint of the old? When personal ego and pride obstruct the way, what room is there for dispassionate enquiry?

So, what has happened is that causal closure of the physical world has spilled over into an attitude of 'mind closure'. Open-ended questioning is rarely welcomed in *orthodox* scientific or medical circles. Nonetheless, international organisations such as the Scientific and Medical Network are fine examples of eminent scientists who champion the rigour and impartial attitude of science but seek to free scientific enquiry from militants who use scepticism as a weapon to defend an ideology or belief system, hence inhibiting the spirit of enquiry.

The Network 'takes the view that the materialistic explanation for the nature of reality is only one perspective [and] follows the guidelines of respect and open mindedness within a framework of academic rigour.'[29] Another such body is Skeptical About Skeptics which 'looks at ways in which scientific objectivity is compromised by vested interests, fraud, experimenter effects, and merchants of doubt who use scepticism as a weapon to further corporate interest.' Both of these organisations passionately uphold science but not scientific fundamentalism. They recognise that healthy doubt and scepticism are important parts of scientific enquiry, but maintain that dogmatic, ill-informed attacks, levelled by self-styled sceptics (especially those who court the popular media), seriously obstruct pioneering scientific research, researchers, and the public understanding of science. They both have eminent scientists on their advisory boards.

There are historical reasons for the current prevalence of the mechanistic paradigm of science. From the seventeenth century until now the scientific worldview has prejudicially ignored the spirit aspect. Such materialistic assumptions progressively hardened by degrees into dogmas during the nineteenth century and fused into an ideological belief system that became known as 'scientific materialism' or 'scientism'.

It constitutes part of the revolutionary shift from the earlier philosophy of the Middle Ages where man was regarded as the focus of creation, to the later view about the less central role of humanity in the overall scheme of things.[30]

The late nineteenth and early twentieth centuries were promising times for a rapprochement between science and spirituality. Thereafter, materialism as a legitimate element of scientific thought gradually hardened into an ideology and then into a dogma. Despite the findings of quantum physics this ideology became so dominant in academia and learned societies in the twentieth century, that the majority of scientists unquestioningly believed that it was based on empirical evidence and therefore represented the one and only credible view of the world.

This, in a nutshell, is the historical *raison d'être* for the present mechanistic and mechanical thinking in science where the whole must be explained by the motion and properties of its physical parts exactly like a clock or a machine. This is reductionism. The near-total dominance of the philosophy and methodology of materialism in academia and industry, medical science and social values has seriously fettered the scientific investigation of mind, consciousness, and spirituality. Such ignoring, or brushing aside, of the inner and subjective dimension of human experience has led to a highly impoverished, distorted and one-sided understanding of ourselves, our place in society and how we relate to the overall scheme of nature.

The mainstream scientific view of the causation of life and mind is strictly one-way upwards: matter → life → brain = mind; i.e., inorganic matter is the primary. Life therefore materializes from matter and brain, which is synonymous with mind, and emerges as the product of life. In this view, the death of the brain means the extinction of the life of the person. There is no question of after-death states or the survival of consciousness beyond the brain.

Limitations in the Modern Scientific Picture

Predominantly left-brained academicians seem unwilling, or unable, to grasp the fact that one cannot experience higher dimensions of consciousness without the necessary altered states of consciousness. One might as well try to view the world from the top of Mount Everest by standing at the foot of the mountain. Essentially, we are blind to that which we are not ready to perceive.

In one sense, the ultra-materialistic stance of science finds its justification amongst those who only look to the outward appearance of nature and choose to deny or ignore her inner working principles. Such outward appearances, however, are always illusory on the physical plane and there is no central, organising principle to synthesise the myriads of theories and conjectures that are put forward in an attempt to explain the nature of the world or to account for mind and consciousness.

So, with all this confusion, is it not reasonable to seek a way out of the impasse? We intend to accomplish this by drawing upon Einstein's counsel: 'We cannot solve our problems with the same level of thinking that created them.'[31] We therefore seek a resolution at a higher level than what purely materialistic science can provide. And that way out is onwards and upwards through the gateway of the eternal wisdom that goes well beyond, but never denies, the physical realm. There is no question about the success of the materialistic philosophy in bringing about greater understanding of the physical world and our physical bodies and brains as well as bequeathing freedom through advances in technology, whilst also admitting the attendant penalty of enslavement to the latter.

Blind faith in 'scientism', as an exclusive explanatory framework for understanding reality has seriously hindered investigation into the subjective and spiritual dimensions of the universe and human experience, thereby diminishing our understanding of our place in nature and of ourselves—who, or what we *truly* are.

We See What We Believe and What We Expect and Want to See

'Turning a blind eye' is an idiom familiar to most of us. We conveniently ignore information that is not compatible with the way we believe things should be, in other words, information that is not sympathetic to our mindset. But there is yet more profound blindness whereby we do not choose to turn a blind eye but literally do not, and cannot, see the truth—it is outside our experience because we have not yet developed faculties by which to see it. There are numerous instances of the former but an anecdote which illustrates the latter is the so called 'missing ships' phenomenon. When the British explorer Captain James Cook arrived in Australia in 1770 aboard HMS *Endeavour*, it was reported that the natives of that land were oblivious of its presence. It has been conjectured that they did not see the ship because their eyes were closed to the possibility of its existence.

Perhaps the following scientific experiment makes the point. In the early 1970s during a talk on paranormal topics as part of a University Conversazione, Arthur Ellison (1920–2000), Professor of Electrical and Electronic Engineering at London University and twice President of The British Society for Psychical Research, suggested to the audience that they try to levitate a bowl of flowers on the lecture table. This was to be done by imagination and concentration and also by chanting the word 'OM' believed in the East to be a sacred sound. After a minute or so of such concentration and chanting the bowl of flowers wobbled, levitated about twenty millimetres into the air, remained there for a few seconds and then crashed down again.

There was a gasp from the audience as they rushed forward to examine the table and bowl of flowers. They could find no normal means or explanation for what they had just witnessed. However, one lady said she noticed a 'greyish substance under the bowl lifting it and the same greyish substance under the legs of the table lifting that.' Then a professor of physics in the audience standing on the edge of the crowd could stand it no longer. So, he pronounced, 'Well, I saw nothing!', turned on his heels and left the room.

Nobody, least of all the lady and the physics professor, was aware that the actual levitation was caused by a purely physical arrangement of an electromagnetic levitator artfully concealed in the drawer of the table and a disc of aluminium stuck to the bottom of the very light plastic flower bowl. The OM chant (amplified through two loudspeakers) was suggested purely in order to hide the sound from the levitator.[32]

Neither the lady nor the physics professor observed the true, objective fact of the actual levitation of the flower bowl but rather 'saw' what they both *believed* and *expected* to see according to their own subjective states, conditioned by their own cultural make-up and belief systems. The lady was presumably a Spiritualist and therefore interpreted the greyish substance as 'ectoplasm' lifting both the bowl and the table. In contrast, the physics professor saw nothing at all since his materialistic outlook—his mental model—admitted of no 'silly nonsense' like levitation by imagination, concentration and chanting 'OM'.

The corollary to this is that we also hear what we want and expect to hear but remain oblivious to what we do not wish to hear. Those who most need to wake up to alternatives hardly ever do. The psychological defences to their paradigms are well-nigh impregnable. Informational inertia, i.e., the aversion to ambiguity and new ideas, is becoming a serious problem!

Vitalist and Organismic Theories

Vitalist and organismic theories argue that evolutionary innovations cannot be explained entirely in terms of chance occurrences or random influences, but must invoke an inner organising, non-mechanistic creative principle, still unrecognised by science. These ideas are completely alien to modern science as they imply purpose-driven design in the material world, such that phenomena are explainable by the overriding purpose that they serve, rather than by the known laws of science. By this premise, evolution does not take place solely on account of spontaneous mutation and natural selection, but rather it is based on the outcome of the purpose towards which the organism is intended to live and function.

Entelechy concerns the essential nature, or inner guiding principle, of a living organism that enables the latter to realise its true potential. It is a goal-orientated factor that directs the system under its control towards its intended aim. Hence, if the normal line of development is disturbed by environmental or any other factors, entelechy re-directs the system towards its internalised purpose along different pathways. But if entelechy is non-physical then how can it bring about physical results?

According to mechanistic theories, memories in humans are stored within the brain cells acting rather like the silicon-chip memory banks of a computer. But there is persuasive evidence to demonstrate that this simplistic outlook is far from the whole truth and that, in fact, memories are not stored physically within the brain and that no satisfactory mechanistic explanation of memory, in terms of physical phenomena, is possible. Near-death and out-of-the-body experiences point to veridical experience (i.e., coinciding with reality) by the subject, involving the retention of the memory of specific events, times, and locations, despite the clinical death of the brain (through accident, stroke or other causes).

Therefore, mind and conscious awareness need not be restricted solely to a functioning brain and may even transcend the brain. This being so, certain kinds of memory need not necessarily be confined to individual minds and memories could therefore pass from person to person, or many individuals could inherit a pooled memory from countless individuals in the past.

The theory of a collective unconscious by Carl Jung (1875–1961) refers to structures of the unconscious mind that are not shaped by personal experience but comprise archetypal forms which are shared among beings of the same species. These archetypal forms could then be regarded as a kind of collective memory pool.

Morphogenesis and Morphic Resonance

The word 'morphic' in the above title comes from the Greek *morphē*, 'form'. Hence morphogenesis is concerned with the question of formative causation: what causes simple organisms to develop in space, and over time, into complex structures having a shape and pattern that uniquely characterises them, for example, an oak tree that grows from an acorn.

We now turn to the work of Rupert Sheldrake (*b.*1942), the internationally acclaimed biologist, previously a Fellow of Clare College, Cambridge, and Research Fellow of the Royal Society. His crowning achievement was an innovative and modernised version of morphogenesis that has greatly contributed towards resolving the shortcomings in materialistic molecular theories of biology. Central to his hypothesis is the idea of morphogenetic fields and morphic resonance.[33]

A 'field' is defined as regions of influence that stretch out beyond the entity from which it radiates. Familiar examples are magnetic, electric or gravitational fields. However, morphic fields differ from the known fields of physics in some important respects. The latter are local to the entity from which they emanate (like the magnetic field surrounding a bar magnet) and extend and connect things in local space.

Morphic fields extend and connect things in both space and time, as well as displaying a universal characteristic, in that they seem to involve not just living systems (biological entities) but also non-biological entities like chemicals and minerals. Furthermore, morphic fields are thought to be non-local—

an important characteristic that has a significant bearing on explaining the nature of instinct and memory and accounting for such phenomena as telepathy.

Morphic resonance is the hypothesis that there is a kind of inherent memory in nature. For example, within a species (or self-organising system), each individual draws upon a collective memory pool and in turn contributes to it. Just as a magnetic field will attract iron filings within its domain of influence, so a morphic attractor will attract developing systems towards it. The hypothesis, backed by experimental evidence, predicts that, for example, new chemicals should get easier to crystallise as time goes on because the crystal forms become increasingly habitual, sustained by morphic resonance from increasing numbers of previous crystals of that type. Likewise, if animals, such as rats, learn a new trick in one place, rats all over the world should be able to learn it more quickly.[34]

The hypothesis of formative causation is supported by impressive evidence for phenomena in minerals, plants, animals, and humans, hitherto unexplained by mainstream science.

The mechanistic approach has not been able to explain satisfactorily some of the central problems in biology such as the development of form, pattern and shape in minerals, plants, animals, and man. The mechanistic-reductionist approach assumes that these must somehow be fully explainable in terms of the complex physico-chemical interactions between the component parts. But this has not adequately explained how increasingly complex structures can grow from less complex ones, for example, an embryo develops from a relatively simple egg cell into a living entity displaying great structural form and complexity.

The development of complex structures progressively in space and time cannot be satisfactorily explained by an exclusively micro-level focus upon interactions between its basic constituent chemicals. It is important to stress that morphogenesis provides the holistic, or organismic complement to the mechanistic and reductionist theory, which partially explains the development of form in space, but not in time.

Inheritance of Acquired Characteristics

The inheritance of acquired characteristics means that, on average, anything that is learnt, or new skills acquired will be progressively easier for succeeding generations of the same species to learn than was possible for their ancestors. Such improved learning is not necessarily dependent upon specifically modified genes being passed on from trained parents to their offspring. This was proved by experiments on rats conducted by the British psychologist William McDougall FRS (1871–1938) in 1920, which showed the same improvement in a specific learning skill displayed by successive generations of rats, descended from both trained and untrained ancestors.

Memory and instinct are thorny problems that modern science cannot fully explain in solely mechanistic terms. The established contemporary view, strongly upheld in mainstream neuroscience, is that memory is stored in the brain in specific 'memory banks', rather like on the hard disk of a computer, and that mind and brain are equivalent terms. The primary evidence proposed in favour of this theory is that brain damage causes loss of memory (analogous to crashing a computer hard disk) and electrical stimulation of the temporal lobes of the brain[35] causes buried memories to be evoked (analogous to powering up a computer to retrieve the data on the hard disk). However, attempts to localise memory traces within the brain have been singularly unsuccessful.

According to Sheldrake, the hypothesis of morphic resonance leads to a radically new concept of memory storage in the brain and of biological inheritance. Memory need not be stored in material traces inside since brains are more like television receivers than video recorders. And biological inheritance need not all be coded in the genes, or their modifications since much of it depends on morphic resonance from previous members of the species. Genes, therefore, act more like transformers and transducers rather than as originators of inheritance and species memory. Thus, everyone inherits a collective memory from the experiences of past members of the species, and also contributes to the collective memory, affecting other members of the species in the future.[36]

Animals inherit the successful habits of their species as instincts because the more often they are repeated, the more probable they become. Humans inherit bodily, emotional, mental, and cultural habits, including the habits of language.[37]

Group Memory

Social groups are likewise organised by fields, as in schools of fish and flocks of birds. Human societies have memories that are transmitted through the culture of the group and are most explicitly communicated through the ritual re-enactment of a founding story or myth. This is especially noticeable in aboriginal and tribal societies characterised by a group memory which individual members tune into. Scientists and anthropologists are beginning to unearth a wellspring of knowledge buried in the ancient stories of Australia's Aboriginal peoples[38] in 'Dream Time', a commonly used term for describing important features of Aboriginal spiritual beliefs and existence that is not generally well understood by non-indigenous people.[39]

Such group memory is also evident in mob violence, racial riots, or gang hooliganism where the people involved behave in ways that would be unlikely when on their own and not part of the crowd. Note that the morphic fields of social groups connect together members of the group even when they are many miles apart and provide channels of communication through which organisms can stay in touch at a distance. They help provide an explanation for telepathy.

There is now robust evidence that many species of animals are telepathic, and telepathy seems to be a normal means of animal communication, as discussed in Sheldrake's book, *Dogs That Know When Their Owners are Coming Home.*[40] Telepathy is normal, not paranormal, natural not supernatural, and is also common between people, especially people who know each other well.

Scientific Breakthroughs

Another area in which the hypothesis of morphic resonance might cast some light is the revolutionary developments in science. There are a number of historical accounts which seem to demonstrate that 'an idea whose time has come' is 'received', almost out of the ether, by several individuals simultaneously. Moreover, it can also be seen that an idea or discovery is brought to its full maturity by several individuals all working at different times, perhaps even spanning different centuries.

The mathematical calculus is a good case in point. The pioneers of the calculus stretch far back to the time of the Babylonians, the Egyptians and the Greeks.[41] During the Age of Enlightenment (Age of Reason) several mathematicians in England and Europe were fervently working on mathematical problems of this genre. However, the complete formulation of the calculus was the supreme achievement of Sir Issac Newton and Gottfried Leibniz (1646–1716), independently and practically contemporaneously.[42]

Organising Fields

A closely similar concept to morphogenetic fields is 'organizing fields'. Gustaf Strömberg (1882–1962) was a Swedish-American astronomer who was a member of the staff at Mount Wilson Observatory from 1916 to 1937. His main work *The Soul of the Universe* alludes to the later discoveries of quantum science, while also supporting the fundamental tenets of esotericism.[43]

By applying the principles of relativity and quantum theory to the field of biology and the relationship between mind and matter, he discovered, 'that the individual memory is probably indestructible, and that the essence of all living elements is probably immortal. The study leads to the inevitable conclusion that there exists a World Soul or God.'[44]

He found our familiar world of space and time, the training school of souls, to be different from the more real 'world' of life and consciousness. There are elements, both material and 'immaterial', which connect these worlds and provide a rational foundation for the existence of the soul and its survival after bodily death. All this is in perfect harmony with the perennial wisdom teaching that no single point in the universe is devoid of life and consciousness of some order. Then comes the most important factor in Strömberg's argument, the so-called 'immaterial wave-structure' in the cosmos which greatly helps to solve the problem of memory, and therefore of survival after death.

The matter in our brains is constantly being replaced, yet memories are accumulated and preserved during a long lifetime. Even long-forgotten events may suddenly flash into vivid consciousness when some inner contact is made. These long-forgotten events also erupt into the mind when there is some pressing *need*, perhaps prompted by external circumstances—as if the urgency triggers the memory which provides just what is needed. How could this be unless an immaterial living structure subsists, in which memory inheres independently of the physical atoms which are changing innumerable times over the course of a lifetime? Furthermore, there seems to be no valid reason why this immaterial structure should not continue without impairment after death, regardless of the dissolution of the physical structure.

Morphogenesis, organising fields, and immaterial wave-structures collectively show the indispensable non-physical, non-local functioning of mind in addition to the working of the physical body and brain. There is a strong allusion to subtle bodies and subtle energies that, in addition to the familiar physical body, constitute our whole being.

Psychology and Parapsychology

Psychological studies have confirmed what is popularly known as 'mind over matter', namely that mental activity can causally influence behaviour, the more conscious and focussed the attention, the more pronounced are the results. It means that mental events can significantly influence the activity of the brain. Put simplistically, mind affects matter, but matter also affects mind: the mental affects the brain but the brain also affects the mental.

Further evidence accrues from psychotherapy, neuroimaging studies and psychoneuroimmunology which collectively show that our thoughts and emotions, expressed in such things as our expectations, beliefs and wishes can bring about changes in our external environment. They can also markedly affect the activity of the physiological systems connected with our brain (such as the cardiovascular, endocrine, and immune systems).[45]

The 'experimenter effect' otherwise known as 'subject-expectancy effect' is a well-known occurrence in parapsychology as well as in scientific experiments or medical treatments. It is a phenomenon that occurs when a researcher expects a given result and therefore unconsciously affects the outcome or reports the expected result.[46] The effect is most commonly found in medicine when a patient is administered a harmless substance but experiences the placebo reaction or nocebo (i.e., negative) reaction, depending on the state of mind of the patient or the expectation of the doctor. Because this effect can significantly bias the results of experiments (especially on human subjects), double-blind methodology is used in an attempt to eliminate the bias effect.

This is closely associated with 'confirmation bias', also known as 'confirmatory bias', which is *an irrational tendency to search for, interpret or recall information in a way that supposedly confirms preconceptions, prior assumptions or hypotheses,* while giving disproportionately less consideration to alternative possibilities.[47] The effect is stronger for emotionally charged issues and for emotional people, especially regarding deeply entrenched beliefs (invariably of a scientific or religious nature). As a result, the experimenter can interpret ambiguous evidence as supporting his existing position because of the tendency to look for information that conforms to his preconceptions and overlooks information that argues against it.

Paranormal and '*psi*' Phenomena – Robust and Repeatable Evidence?

Psi is a collective and generalising term for the aggregate of parapsychological functions of the mind including extrasensory perception, precognition and psychokinesis (the psychic ability to influence a physical system without physical interaction). Research into *psi* phenomena indicates that living organisms (including human beings) and physical apparatus can be influenced mentally. Possibly, the most astonishing evidence of psychokinesis ('PK'), the power of the human mind to exert non-physical control over inanimate objects, comes from the American plasma physicist Robert Jahn (1930–2017) and colleagues of the Princeton Engineering Anomalies Research (PEAR) program.[48]

Millions of electronic, mechanical, optical, acoustic, and fluidic experiments were performed over almost a quarter of a century to reveal incontrovertible evidence of the power of 'mind over matter'. All these devices produced strictly random outputs when unattended and were calibrated precisely to establish their unattended statistical output distributions. The human subjects proved capable of 'mind-altering' the output of the devices, such that the chances of obtaining such a bias by coincidence alone was calculated to be less than one in one thousand billion.

The Princeton experiments have shown some astonishing facts that have been admitted even in 'hard' science journals. First, that distance and time do not figure in the results.

People thousands of miles away seem able to affect the machines as if they were in the same room. Then, operators do not even have to apply their thoughts whilst the device is operating. Another characteristic is that each individual seems to have a unique effect on the machines, producing in effect his personal 'signature'.

Predictably, the fact that the detailed output of this daring research, conducted using the utmost scientific rigour, was never mentioned in the backlash against Jahn and the PEAR team was because, as pithily stated in 2003 by Adrian Berry, Science Correspondent of the British national newspaper *The Daily Telegraph*: 'Few subjects more infuriate scientists than claims of paranormal phenomena, because if confirmed, the whole fabric of science would be threatened.'

Dean Radin (*b.*1952) is one of the foremost researchers of *psi* phenomena at the Institute of Noetic Sciences (IONS). Using meta-analysis (combining thousands of results over hundreds of experiments covering a century of increasingly sophisticated research), Radin has demonstrated, conclusively, a range of *psi* phenomena and abilities. These experiments can be repeated to the same effect. The existence of telepathy (mind-to-mind perception), clairvoyance (perception at distance), precognition (perception through time), psychokinesis (mind-matter interaction) and other *psi* phenomena (e.g. mental interactions with living organisms) are all shown to be incontrovertible. This book definitively lays to rest any doubts as to the experimentally demonstrated existence of at least some psychic (or *psi*) phenomena.

Near Death Experiences (NDEs)

Advances in medical science have demonstrated that a person pronounced clinically dead (no breathing, no heartbeat, no electrical brain activity registered on the electrodes of the electroencephalograph machine for a few minutes or sometimes even longer) may be revived and can sometimes describe in vivid and accurate detail the phenomenon known as a near death experience NDE. This has shown that conscious mental activity and memory can be experienced even when there are no vital signs of physical life during a critical accident or cardiac arrest.

What does the Perennial Philosophy have to say about NDEs? The *Tibetan Book of the Dead*[49] is intended as a guide through the experiences of consciousness after death so that the dying person may recognise the nature of his mind in the bardo, the interval between death and the next rebirth. The Tibetan book has a major bearing on the NDE because it teaches that once awareness is freed from the body, it creates its own reality as one would experience in a dream.[50] This dream occurs in various phases in ways that are wonderful or terrifying, with peaceful or wrathful visions occurring. This shows a similar pattern of events to the NDE except that in an NDE such disconnection with the physical body is temporary as the experience is one of near-death and not death itself.

The journey of consciousness after death, and in the afterlife, is also described in detail in the *Egyptian Book of the Dead*[51] and in the legend of the 'Myth of Er' that concludes Plato's *Republic*.[52]

Out-of-Body Experiences (OBEs)

OBE is a modern term for 'astral projection' used in the occult literature mainly of the East. OBE subjects state that they seem to possess what may generally be described as a 'second body' moulded by the influence of thought, so much so that thought and action seem to be much more closely coupled than in normal life. Subjects also identify different kinds of space in which they find themselves.

Sceptics should note, carefully, that it is entirely possible to train oneself to have an OBE.[53, 54, 55] The essential message from OBEs is that quite clearly there can be more than one stream of consciousness operating at one time. Given that consciousness needs an appropriate vehicle through which to manifest and express, this implies that there can be more than one 'body' used at one time—bodies other than the physical, namely, the subtle bodies.

Lucid Dreaming

The lucid dreaming experience is closely linked to the OBE. It can be induced using various techniques or it can occur during sleep. The eminent British scientist and Theosophist E. Lester Smith FRS (1904–1992) suggests that our dreams can loosely be grouped into four categories that are by no means watertight.[56] The common dream is one where we see the various scenes and episodes as passive observers. In other types, we ourselves seem to play a role but one that is imposed on us as if we were actors in a play written by another playwright. In the third type of dream, we seem to play a more active role. Whereas the dream scenery is invariably imposed, we can choose what we do in it. This is the so-called lucid dream in which we are aware that we are dreaming and can deliberately take control over the course of the dream events. There is also a fourth kind of dream where a few people have been able to pre-programme their dreams and decide, in advance, to visit a place or a particular person.

Mediumship

Mediumship is generally understood to be the ability of a person to communicate with someone who has physically died. On what level such communication occurs is another highly complex matter that will not be considered here. Nonetheless, there is documented evidence from controlled laboratory experiments that skilled research mediums can sometimes obtain specific and detailed information about deceased persons.

Arthur Ellison participated in numerous séances and meticulously investigated dozens of mediums all throughout his life. He stresses that, contrary to public perception, mediums are amongst the most honest and genuine people he has known and that one very rarely meets a fraudulent medium. This testimonial, coming from an impeccable scientist and lifelong psychic investigator, flies in the face of unsubstantiated, throwaway opinions from scientists who have never bothered to investigate such phenomena.

In essence, mediumship, like all the other phenomena described, provides further evidence about consciousness not necessarily restricted to the physical brain—in other words, that mind can exist apart and separate from the brain.

Remote Viewing and Remote Dowsing

Obtaining an image of a distant place existing in the present by focussing the attention on it is known as 'remote viewing'. It uses the same core faculty as Dowsing, but in different ways. The distant location need not necessarily be on planet Earth. For example, the English engineer and consciousness researcher Peter Stewart FREng (1928–2018) used remote dowsing on a map of the lunar surface when he carried out a feasibility study for the Hawker Siddeley Aeronautics Group for 'A Programme for the surface exploration of the Moon'.[57] Another good example is the remote viewing of the planet Jupiter by the American psychic Ingo Swann (1933–2013), years before the planet was explored by the NASA/JPL space probe, which confirmed the accuracy of the viewing.[58]

Unsurprisingly, the major research effort into remote viewing has been for military intelligence, such as the remote viewing at Stanford Research Institute initiated by the US Central Intelligence Agency (Stargate Project).[59] Jessica Utts (*b*.1952), an American parapsychologist and statistics professor at the University of California, confirmed the statistical likelihood of remote viewing and precognition as having real effects: 'For many years I have worked with researchers doing very careful work in this area, including a year I spent working on a classified project for the United States government, to see if we could use these abilities for intelligence gathering during the Cold War. At the end of that project, I wrote a report for Congress, stating what I still think is true. The data in support of precognition and possibly other related phenomena are quite strong statistically, and would be widely accepted if they pertained to something more mundane.'[60]

Remote viewing can also be done in time so that, intriguingly, information can be obtained about a distant place in the past, but no longer existing or about a place or event that will materialise in the future. The most outstanding example of the use of this faculty was when one of the United States Army remote viewers, Joe McMoneagle (*b.*1946) remote-viewed a new Soviet nuclear-powered submarine of the 'Kursk' class and was then able to 'fast forward' into the future to ascertain the launch date. The US 'KH9' spy satellite was positioned over the Soviet shipyard and recorded the launch of the submarine on the date predicted by McMoneagle.

Remote viewing, then, is not the province of the gullible or overly credulous, albeit such types abound in this as in other areas involving altered states of consciousness. It has shown the non-local, non-temporal nature of mind plus its ability to access past and future events in our ordinary space-time world. From the standpoint of this book, remote viewing clearly shows that consciousness is not limited to the brain and extends far beyond it.

Quantum Physics Cracks the Backbone of Materialistic Science

The onset of quantum physics was indeed a unique and spectacular event in the annals of science. It may, for good reason, be regarded as the crown jewel of science, for it, above all other scientific disciplines, has cracked the hitherto exclusively materialistic paradigm of science and implied a non-material universe where consciousness need not be regarded as the product of solely material processes.

The process of scientific awakening that was to undermine the materialistic foundation of classical physics was occasioned by five events around the turn of the twentieth century: (*1*) 1897—the discovery of the electron by J. J. Thomson (1856–1940); (2) 1898—the discovery of radium by Marie (1867–1934) and Pierre (1859–1906) Curie; (3) 1900—Planck's discovery of quanta; (4) 1905—Einstein's famous equation showing that energy and matter are convertible; and (5) 1911—the discovery by Ernest Rutherford (1871–1937) that atoms are practically 'empty space'.

The Five Cornerstones of Quantum Physics

1. *Indivisibility of quantum action*: a system changing from one state to another does so in discrete transitions ('quantum leaps') and not in a smoothly continuous series of intermediate states, as believed in classical physics.
2. *Probabilities rather than actualities*: such that every physical situation is characterised by statistically formulated potentialities rather than actual properties as held by classical physics.
3. *The uncertainty principle*: as formulated by Werner Heisenberg, position and momentum or energy and time cannot be determined with equal accuracy. Accurate determination of the position of a particle alters its momentum with consequent uncertainty in the value of momentum.
4. *Instantaneous action-at-a-distance*: paired particles apparently disconnected in space can communicate instantaneously.

5. *Particle ~ Wave Duality*: It is now universally acknowledged in modern science that waves and particles are two aspects of one reality. Every entity in nature presents dual wave and particle characteristics.

Quantum physics has undermined the material basis of our worldview that subatomic particles and atoms are solid entities existing with absolute certainty at a precise spatial location and a definite time. Crucially, mind has entered the picture, such that for many quantum physicists the consciousness of the observer is a vital component to making sense of the physical events being observed. These results strongly imply that the world of physical matter is neither the fundamental nor the sole component of reality and indeed (as unequivocally affirmed in the Perennial Philosophy) that mind is primary.

Occult Chemistry

Occult Chemistry refers to a book by that name about a special case of remote viewing or clairvoyance, namely, that of subatomic particles.[61] Experiments were conducted at intervals over a period of nearly forty years, from 1895 to 1933, by two pioneering British Theosophists, Charles Webster Leadbeater (1854–1934) and Annie Besant using a form of extrasensory perception virtually unknown to parapsychologists, although long known in India as a yogic *siddhi* (psychic faculty) having the Sanskrit name of 'anima'.

What was observed, and meticulously recorded by them in *Occult Chemistry*, has demonstrable consistency with the established facts of nuclear physics, quarks and superstring theories. (Quarks are any of a number of types of subatomic particles and a fundamental constituent of matter. Superstring theory attempts to explain all of the particles and fundamental forces of nature in one theory by modelling them as vibrations of tiny supersymmetric strings.)

Not all their interpretations can be taken at face value because of the 'observer effect', namely, the disturbance of the quantum dynamics of particles by the act of clairvoyant observation (a possible manifestation of Heisenberg's Uncertainly Principle). Nonetheless, this is the only recorded example of a whole series of extrasensory perception (ESP) observations where one can have cast-iron certainty that any charge of fraud, or the unconscious use of sensory clues, can be excluded. This is because of the total absence of any scientific information at the time on what they foresaw and verified and which was only discovered decades later.

For example, the first edition of *Occult Chemistry* appeared in 1908. The British Nobel physicist Ernest Rutherford PRS (1871–1937) confirmed the nuclear model of the atom two years later. The neon-22 isotope—clairvoyantly detected in 1908—was discovered four years later. The Danish Nobel physicist Niels Bohr FRS (1885–1962) presented his theory of the hydrogen atom five years later. The British Nobel physicist Sir James Chadwick FRS (1891–1974) discovered the neutron twenty-four years later. And the first quark model was proposed fifty-six years later in 1964 by the American Nobel physicist Murray Gell-Mann FRS (1929–2019) and George Zweig (*b.*1937).

After making appropriate adjustments for the observer effect, the numerous diagrams, figures and descriptions of the chemical elements in *Occult Chemistry* have shown the clairvoyant viewing of atomic particles to be largely corroborated by modern particle physics as shown by the Cambridge-trained English physicist Stephen Phillips (*b.*1946).[62] Accordingly, it shows the power of consciousness to access information in subatomic realms. This makes a substantial case for extended mind; hence consciousness not being generated by, or the product of, material processes in the brain.

Reincarnation

By 'reincarnation' we mean the repetitive re-embodiment of the human soul in bodies that are physical, spiritual and ethereal or subtle. The 'Human Soul' is the 'individuality' of man—immortal in essence, as distinct from his mortal personality which includes body and brain. It cannot be emphasised too strongly that the personality, body and brain permanently disintegrate upon physical death and do not reincarnate. It is man's higher principle, the individuality, that 're-infleshes' by assimilating a new personality conditioned by karmic law. This is the law of cause and effect operating on all levels— physical and moral—whereby a man reaps what he has sowed, whether in this life or previous lives.

Belief in reincarnation is universal amongst many ancient philosophies and religions of both East and West. It was taught in the esoteric schools of India, Egypt, Greece and elsewhere. The doctrine is universally propounded by enlightened philosophers, sages, and saints ancient or modern.

A reader who desires just one volume as a standard reference can do no better than to peruse *Reincarnation: The Phoenix Fire Mystery* by the American Theosophical author Sylvia Cranston (1915–2000).[63] This fabulous work is an anthology in the true sense of the term, meaning a symphony of ideas, about death and rebirth garnered from the worlds of science, religion, philosophy and psychology and also from history, mythology, art and literature. Suffice it to say that this vast body of material and cited evidence allude strongly to realms of existence that are not based solely on physical matter. They also fly in the face of a purely materialistic science that equates the extinction of life with the death of the brain.

Our science-dominated Western culture has, on the whole, jettisoned any notion of reincarnation, pronouncing such ideas as 'unscientific' or 'pseudoscience'. But such is not the case with other cultures, especially in ancient Asia, including of course

India. It is the arcane sciences of both the West, and especially the East, that have provided chapter and verse on a science of the subjective which recognises the primacy of consciousness. In the West, there is now a slow but inescapable recognition of a totally new science of consciousness.

Arguably, the finest contribution to scientific reincarnation research in the West was done by the Canadian-born U.S. psychiatrist Ian Stevenson (1918–2007) of the University of Virginia. He became internationally recognised for his research into reincarnation (from a non-religious perspective) by discovering evidence from children suggesting that memories and physical injuries can be transferred from one lifetime to another. His meticulous research presented evidence that such children had unusual abilities, illnesses, phobias and affections, which could not be explained by the environment or heredity.[64] He travelled extensively over a period of some forty years, investigating three thousand cases of children around the world who recalled having past lives. His investigations outside of Western culture have been recorded in four volumes of his work bearing the overall title of *Cases of the Reincarnation Type*.[65]

Stevenson placed much emphasis on children exhibiting birthmarks whose location coincided with the recollected site of a claimed past-life lethal injury.

Loss of past-life memories is often used as an argument against reincarnation. In humans there is the phenomenon of *neonatal amnesia* whereby past-life memories in children are more difficult to access after five to seven years of age, a time invariably concurrent with the rapid onset of neurogenesis where new neurons are formed in the brain.

Much evidence about reincarnation also comes from spiritualism and it is not all a case of smoke and mirrors, delusion and gullibility. Certainly, this area has attracted its fair share of charlatanry but there is also testimony from the best side of spiritualism that has attracted serious attention from eminent scientists in the nineteenth century and in modern times.

The finest aspect of spiritualism is arguably the rescue circle, which brings in other interesting facets of life after physical death, such as earth-bound souls, apparitions seen by several people, premonitions, and phenomena of rapport.[66] All of these cases undermine a purely materialistic basis of life, which is why we place such emphasis upon them. Spiritualism then, was helpful in countering the deeply materialistic thinking of the time and the existence of normally unseen worlds came to be recognised and accepted by science. But as is the nature of things, the spiritualist movement gradually lost its scientific investigative spirit and took on, by degrees, the tenor of religiosity.

Carl Jung stated in his book on archetypes[67] that rebirth and reincarnation must be counted among the primordial affirmations of mankind which are more powerful than the ill-researched and unsubstantiated assertions of any number of sceptics who doggedly refuse to look at the weight of evidence before their eyes. However stated, in whatever language, mankind has affirmed the existence of invisible realms of being that simply cannot be accounted for on the basis of physical science and medicine alone.

Such invisible realms imply grades of substance finer than physical matter and subtle (non-physical) bodies as the appropriate vehicles of consciousness. The huge weight of evidence from diverse sources in favour of reincarnation strongly suggests that scientists should adhere to the spirit of science and be prepared to modify their materialistic paradigm and mechanistic theories to match the evidence. They should not dismiss the latter on the grounds that a satisfactory theoretical model and explanatory framework do not exist.

'Extraordinary claims require extraordinary evidence' was a phrase coined by the American astronomer, astrophysicist and cosmologist Carl Sagan (1934–1996). It has now become virtually a universal mantra embedded in the scientific ethos and proffered as a key issue for critical thinking, rational thought and scepticism for what is regarded as weak evidence. What counts as 'extraordinary evidence' will always remain ill-defined. The production of electricity, the electric light bulb, the fact that the behaviour of subatomic particles would depend on conscious observation, the penetrating power of x-rays, the splitting of the atom and fission of slightly less than sixty-four kilograms of Uranium-235 releasing energy equivalent to approximately 16,000 tons of TNT (as in the Hiroshima atomic bomb), and heavier-than-air flying machines, are just a few a examples of what would have been extraordinary claims at the time but now, in the light of overwhelming evidence, seem commonplace and anything but extraordinary.

A spectacular U-turn in science in recent years concerns the condition of the planet Mars. At one time the Red Planet was unequivocally pronounced by astronomers to be dry, barren and dead. In recent years, NASA missions and further astronomical observations have shown the presence of distinct polar ice caps.

In neuroscience the firmly established concept that the brain is a physiologically static organ has been displaced by the discovery of neuroplasticity, which shows that the brain changes throughout life; and brain functions are not confined to certain fixed locations. This exciting discovery is pursued in greater detail shortly, as it has a major bearing on the whole question of consciousness in mainstream science.

Extraordinary biological evidence has led to the extraordinary claim that genes alone cannot explain the unique characteristics of humans, as was once the dictum. We confidently look forward to the time when 'extraordinary evidence' will pave the way to an outlook on evolution that is primarily rooted in spirituality and consciousness, whilst not ignoring the physical Darwinian theory.

'Panspermia' is a recent hypothesis about life existing throughout the universe, distributed by meteoroids, asteroids, comets, planetoids, and also by spacecraft in the form of unintended contamination by microorganisms. Such microscopic life forms are purported to survive the effects of space, becoming trapped in debris that is ejected into space after collisions between planets and small solar systems that harbour life. This would suggest that the ingredients of life—proteins and amino acids—may have formed in the depths of interstellar dust clouds and that life throughout the galaxy is universal. When the visionary English astronomer Sir Fred Hoyle (1915–2001) first mentioned the idea back in the 1950s, it was regarded as so bizarre that, despite his prestige, no scientific journal would publish his paper.

What science has pronounced as a 'gospel truth' at one time has been overturned by that very same science in the light of its own further discoveries. Slowly and inexorably it points towards the Eternal Wisdom which affirms that even the smallest corner of the universe is teeming with life and consciousness. Furthermore, the crucial role of comets in relation to the evolution of Earth is another important tenet of occult cosmogenesis.[68]

There are bound to be more such U-turns in science concerning extra-terrestrial planets and, indeed, concerning our own planet. The writer confidently foresees an imminent, major turnaround in geology regarding the acceptance by mainstream science of the prehistoric continent of Atlantis and even the continents that preceded it. Whereas every occult teaching, from antiquity to modern times, has declared the existence of great ancient civilizations on the basis of the accumulated wisdom of the ages and the testimony of generations of seers, the corpus of geologists still choose to deny their existence.

Without taking any cognisance of the weight of the occult teachings, their denial is based primarily on plate tectonics, the current theory in geology that the outer shell of the Earth is divided into several plates that glide over the mantle, the rocky inner layer above the core. However, even the geological evidence is by no means cast in stone. There are several anomalous facts and findings alluding to prehistoric continents.[69]

It seems that major U-turns in science are grudgingly at first, then universally, accepted by the scientific community as part and parcel of the onward march and progress of science. But in the broad field of spirituality or psychic investigation, any uncertainties, vague definitions or the smallest deviations from previously held views, are immediately pounced upon by many establishment figures and both the subject, and the researcher, are dubbed 'New Agey', 'woolly', or 'flaky'. Who has granted establishment science immunity from the charge of flakiness?

Enough has been said to demonstrate that it is as absurd, as it is unscientific, to use the Sagan cliché as a weapon to dismiss robust evidence that cannot be explained materialistically or mechanistically. This is not to deny that there are innumerable examples of charlatanry or claims to paranormal feats based on wishful thinking, deception or self-projection. However, spurious or sham claims say a lot about the falsity of the claims and the character of the claimant, but they say nothing whatsoever about independent, genuine evidence.

Perhaps we may add that the *exclusively* materialistic worldview pronounced as an overarching scientific philosophy, rather than applied within its legitimate boundaries, and founded entirely on the assumption of the primacy of dead matter and a godless universe, is a claim for which it is difficult to find evidence, weak or strong, ordinary or extraordinary.

Knowledge Suppression Mechanisms

There is a vast amount of provocative evidence that blatantly contradicts currently accepted ideas in mainstream science. Such evidence in all fields of science has been systematically suppressed, forged, mutilated, ignored, allowed to be forgotten or brushed aside as 'witchcraft' or 'snake-oil'. We can discern two primary suppression mechanisms. The first is the deliberate attempt by a few politically powerful scientists to block any evidence that runs against established opinion (upon which large research grants and industrial funding also depend). The second is more in the nature of an on-going social process of knowledge filtration that appears insignificant in small doses, but has a major cumulative effect over time, whereby any kind of embarrassing evidence is quietly allowed to disappear into oblivion. The first suppression mechanism operates mainly in the fields of biology, cosmology, and physics; and the second in palæoanthropology and evolutionary theory.

A Fine Definition of Pseudoscience

Nowadays, it is certainly the case that a scientist embarking on their career, or even a world-famous scientist who espouses unfashionable ideas and fails to conform, thereby daring to step outside the canons of scientism, is swiftly excommunicated and may well find his career and reputation in shreds. Woe betide a young researcher who wishes to embark on a career in parapsychology! The writer suggests, therefore, that the devotees who worship in the 'Church of Scientism' are, in fact, pseudoscientists.

Earlier we quoted the statistics professor at the University of California, Jessica Utts, confirming the statistical likelihood of remote viewing and precognition. She added: 'When I have given talks on this topic to audiences of statisticians, I show lots of data. Then I ask the audience, which would be more convincing to you—lots more data, or one strong personal experience? Almost without fail, the response is one strong personal experience!'[70]

In view of scientism's exclusive devotion to materialism would it be fair to refer to scientism as a cult? Would it be unreasonable to claim that being inimical to the true ethos of science is fundamentally anti-science, in fact, pseudoscience, and that those who uphold it singularly belong to the cult of scientism?

> *The day science begins to study non-physical phenomena, it will make more progress in one decade than in all the previous centuries of its existence.*
>
> Nikola Tesla[71]

Beyond 'Uncomfortable Science'

History has taught us that even the best theories will ultimately need to be revised and the successful solution to the pressing scientific problems of today cannot therefore be left solely to the custodians of the past but will have to include the visionaries and pioneers of our time, many of whom have been suppressed.

Our conclusion is that the existing, mainstream evolutionary model cannot be adapted or modified but must be jettisoned entirely and replaced by a radical new model without preconceptions and with consciousness as its underlying premise.

We have shown the limitations and deficiencies of modern scientific picture of our world and ourselves. This discomfort with the establishment viewpoint, especially concerning mind and consciousness, has led avant-garde science to embrace a wider and deeper perspective. A more complete and open paradigm is necessitated:

1. The universe is a multi-dimensional, multi-levelled, organic, and living entity;
   ... consequently ...
2. it displays innate intelligence, order, and purpose;
   ... therefore ...
3. why limit knowledge and experience to the physical realm or just the five senses?
   ... So also ...

4. consciousness need not be restricted to physical matter alone or solely to the brain;
   ... in which case ...
5. consciousness may survive the death of the brain;
   ... what is more ...
6. consciousness may even be fundamental rather than a by-product of material interactions;
   ... and in fact ...
7. consciousness may be unqualified and unconditioned and could display a spectrum of states with matter as its effect, rather than its cause.

The chief insights for science, and associated benefits to humanity, of this broader and deeper perspective regarding consciousness are that:

1. Consciousness and mind may represent the primordial aspect of reality and so cannot be a derivative of matter.
2. Mind being apparently unbounded, there may be a unity of minds implying a unitary One Mind which subsumes all individual minds.
3. There may be subtle connections between minds and between minds and the physical world.
4. The brain may act as the transceiver and transducer of consciousness, meaning that mental functions use the brain as an instrument rather than as a generator or producer of thought.
5. The clear implication is of other levels of reality that are non-physical for which non-physical, or subtle, bodies are the appropriate vehicle of expression of consciousness.
6. Focussed, concentrated mind, applied as will or intention, may influence the individual mind and other minds, physical matter and the state of the physical world

not confined to specific spatial locations or to specific moments in time.

7. Given the non-local operation of mind, the expectations (along with the personal emotions, conditioning, and prejudices) of a scientific experimenter may never be completely isolated from the experimental result.

The Present World Situation

Peer pressure and the existing politics of knowledge are still powerful factors impeding open-minded enquiry and scientific progress. Nonetheless, there is no reason for scientists to shirk from open and rigorous research of spirituality and non-physical approaches to consciousness for fear of ridicule from colleagues. An enlarged science embraces and includes, but is not restricted to, a mechanistic science and applauds the scientific achievements that have resulted from empirical observations. By not restricting itself just to matter and physical mechanisms, there will ensue a deeper appreciation, and expanded capacity, to understand the finer and inner workings of nature and man, as well as the pre-eminence of spirit and mind in the unfoldment of the universe.

The logical outcome will be enhanced awareness and sensitivity towards our environment and increased human dignity by virtue of compassion and respect between ourselves, towards all beings, and our whole planet. Let us remember that this expanded view of universe, nature, and mankind was always the bedrock of the Perennial Philosophy, alchemical practices, body-mind-spirit traditions and the esotericism of both East and West. It has just been virtually forgotten during the past four hundred years or so and sorely needs to be recovered and restored to its rightful place.

At the moment, humanity still seems to be working on the physical and objective level. This is because the sensitivity of the average person about the environment only seems to heighten as a result of fear of what might happen if the trajectory of economic growth, combined with environmental degradation, continues its present course. With each natural disaster that fear arises and is compounded. There is now a panic to do something about it. Unfortunately, however, 'fire-fighting' without the enlightenment of the Perennial Philosophy seldom produces a long-lasting and positive outcome.

Signs of Optimism – Post-Materialist Platforms for Science and Consciousness

Despite the current predominance of the machine paradigm, there are encouraging signs ahead. The Galileo Commission is a recent project of the Scientific and Medical Network, one of whose principal aim is to challenge the adequacy of the philosophy of scientific materialism as an exclusive basis for knowledge and values. It consists of a distinguished group of over a hundred scientific advisers affiliated with more than thirty universities worldwide. The Commission Report invites scientists and academics to look dispassionately at such research evidence that is currently ignored or dismissed because it is philosophically incompatible with materialism.[72]

The purpose of the report is to open up public discourse and to find ways to ensure that science is not constrained by an outdated view of the nature of reality and consciousness and is better able to accommodate and explore significant human experiences and questions that it is currently unable to accommodate for philosophical reasons. It anticipates that expanding science will involve new basic assumptions, additional ways of knowing, new rules of evidence and new methodologies.

If science could be based on an expanded set of assumptions, and if they came to form the dominant philosophy of science, then that would open up new avenues and new possibilities and would transform the current worldview constrained by materialism.

Consciousness – an Elastic Term

'Consciousness' is a complex and elusive term with a spectrum of meanings. It is a phenomenon not an object. Consciousness is absolutely central to Eastern (particularly Asiatic) philosophy and thought and is expressed in the Sanskrit tongue in all its subtlety and fine shades of meaning. It is derived from the Latin prefix *con-* 'with, together' and *scīre* '(to) know'. The noun 'consciousness' first appeared in English in the 1630s meaning 'internal knowledge' and, from 1746 it was defined as a 'state of being aware'. In 1863 we move towards the more modern usage when the word 'conscious' was defined in the Oxford English Dictionary as 'having an internal perception of consciousness'. The term 'consciousness raising' was first introduced in 1971.

American psychologist and parapsychologist Charles Tart (*b.*1937) and others have coined the phrase 'altered states of consciousness' to distinguish and differentiate such states from 'normal, everyday consciousness'.[73] These can be attained by abnormal breathing practices, psychedelic drugs or purely mental techniques such as yoga.

Increasingly, nowadays, we find phrases with a self-reflexive content such as 'to become fully conscious entails being conscious of being conscious'. Accordingly, a state of consciousness is characterised by both the nature and the level of its awareness. The whole field of consciousness can, for analytical and explanatory purposes, be stratified into three broad, but highly overlapping 'wavebands'.

Physical Consciousness

The basic level of consciousness is physical brain consciousness which can be likened to a television that is switched on with the electric power and internal circuits functioning but not tuned to a particular frequency, therefore not receiving any signals. All the flow of consciousness is internal. The physical consciousness, therefore, necessarily demands physical approaches to enquiry and investigation.

What is popularly known as 'the unconscious' is the reservoir of feelings, thoughts, urges and memories that lie outside of our conscious awareness. According to the psychoanalytic pioneer Sigmund Freud (1856–1939), the unconscious continues to influence our behaviour and experience, even though we are unaware of these underlying influences. The unconscious may also be those autonomous processes constantly in operation (even during sleep) concerned with the functioning and maintenance of the physical brain and therefore not under the direct control of the individual.

This 'brain-only consciousness' can be regarded as something that can be examined objectively in detachment from the self. Its functions include:

1. Ability to discriminate, categorise, and react to environmental stimuli.
2. integration of information by a cognitive system.
3. reportability of mental states.
4. ability of a system to access its own internal states.
5. focus of attention.
6. deliberate control of behaviour.
7. difference between wakefulness and sleep.[74, 75]

With no stimulus there is a clean 'circuit of consciousness' within the individual. However, the onset of any appropriate stimulus—whether external or internally generated by perception or imagination—may cause the emergence of any one of numerous psychological states that distort the original baseline consciousness. The occurrence, intensity, and distribution of such distortions will vary in individuals and also in groups, depending upon such factors as propaganda, peer group pressure, advertising, manipulation by family and friends and even the time and nature of the environment.

Lower Consciousness

The next level of consciousness is the interface between the physical and the non-physical consciousness. It corresponds to the 'lower mind', which acts as a tuner. This lower consciousness encompasses the physical body and three lower subtle bodies comprising the mortal personality of man (explained more fully later).

Higher Consciousness

This corresponds to the waveband of the 'Higher Mind', acting as a tuner to access information at a higher frequency level. This consciousness encompasses the three higher subtle bodies comprising the immortal individuality of man (also explained more fully later).

The physical body and six subtle bodies (three lower and three higher), and their corresponding centres of consciousness, form what may be described as a spectrum of consciousness of 'ascending frequencies'. It cannot be emphasised too strongly that there is no question of seven separate consciousnesses in impermeable frequency bands. There is but ONE consciousness.

Group Mind

When we consider the high degree of organisation involved in the construction of beehives or ant hills, the obvious deduction is some form of overshadowing consciousness, or group mind, that directs the process. Another familiar example is the beautifully co-ordinated flight patterns of a flock of birds like geese or starlings during their migration. Group consciousness and group mind also exist among humans. As alluded to earlier, it is questionable whether any member of a lynch mob would behave violently in the same way, if left to himself, as he would do when partaking of the 'mob consciousness'.

The intelligentsia have their own version of 'mob rule' as legendary scientists, who dared to step out of line with their peers, have found to the detriment of their careers and reputations. When an idea sweeps through a nation, and sometimes multiple nations, it is as if there is a kind of group consciousness about an 'idea whose time has come'. An example of this was cited earlier concerning scientific breakthroughs regarding the complete formulation of the mathematical calculus independently and practically contemporaneously by Issac Newton and Gottfried Leibniz. Another kind of uplifting group consciousness is seen when people are united in prayer or meditation in a sacred space, whether that be in the countryside, a church, or cathedral. There is also an online group consciousness.

The Swiss psychologist Carl Gustav Jung used the term 'individuation' to describe the process of becoming an individual whereby all potentials become fully conscious and operative. It is a process of psychological integration by which individual beings are formed and differentiated from other human beings and the herd instinct of the crowd. To be aware of one's own individuality in no way implies an isolated or self-enclosed consciousness—individuation being the process by which all that is potential in us, becomes actual and we then become fully ourselves. On the contrary, strong individuals will invariably form, or be drawn to, societies having a common interest or to groups having a common altruistic or charitable motive. The collective consciousness of such societies, characterised by immense individual diversity within a unity of common purpose, is radically different from the mindset of oppressive uniformity and imposed standardisation found amongst totalitarian regimes.

Cosmic consciousness refers to the state whereby, after long processes of self-refinement, the individual consciousness voluntarily becomes fully merged in the universal field of consciousness (known in the East as *samadhi*). It is supposed to mark the full flowering and culmination of the human journey of self-development. As one might expect, this state occurs in various degrees amongst people, or groups, with a strongly religious, mystical, philosophical or artistic bent, but it can also occur amongst ordinary persons. In all cases, the person experiences flashes or extended moments of cosmic awareness. However, even one such genuine momentary occurrence is often enough to alter, radically, the course of the subject's life towards a more philanthropic and less worldly outlook.

The Nature of Memory

Memory is a *generic term* to encompass three faculties—remembrance, recollection and reminiscence. Despite some overlap, they are most certainly not synonyms. It is the innate faculty in thinking beings (including animals) of reproducing past impressions by an association of ideas depending entirely on the functioning of the physical brain. Remembrance and recollection are the attributes of memory. But reminiscence, as understood by modern psychology, is something intermediate between remembrance and recollection, a kind of conscious process of recalling past occurrences but without specific reference to the particular events which characterised the recollection.

We may go further and state that in the light of the Perennial Philosophy, memory is physical and depends on the physiological conditions of the brain; but reminiscence is the longer lasting *memory of the soul*. Memory is a recording device that can easily malfunction but reminiscence is eternal, therefore imperishable. This is based on the profound teaching of Indian philosophy about *Akasha*, the equivalent Western term being *Æther*. As explained by the Hungarian philosopher of science Ervin László (*b.*1932), Akasha is 'an all-encompassing medium that *underlies* all things and *becomes* all things';[76] therefore, it is the womb from which everything manifested has emerged and into which everything will ultimately be reabsorbed. '"The Akashic Record[s]" is the enduring record of all that happens, and has ever happened, in the whole of the universe.'[77] The discoveries and theories of the very latest science are pointing strongly to this profound tenet of Indian philosophy, but it is a new concept in Western science.

In the broadest sense, remembrance, recollection, and reminiscence relate to time in the short, medium, and long term.

A New Continent of Thought

Is it possible to alter radically the way we ordinarily think along well-established tramlines of thought? Can such fixed patterns of thought be broken and transcended? Can our thoughts change the physical structure and functioning of our brains? The answer is a resounding 'yes'. Then, why do the materialists think the way they do, unable to appreciate a wider viewpoint?

Brain plasticity refers to changes in neural pathways and synapses due to changes in behaviour, environment, neural processes, thinking and emotions, as well as changes resulting from bodily injury.[78] Neuroplasticity has replaced the former well-established position that the brain is a physiologically static organ and explores how, and in which ways, the brain changes throughout life.

That the brain and its functions are not fixed throughout adulthood was proposed in 1890 by the psychologist William James, though the idea was largely neglected.[79] It then germinated in 1923 when the psychologist Karl Lashley conducted experiments on rhesus monkeys. By removing certain parts of the brain of these creatures he demonstrated changes in neuronal pathways which he concluded to be evidence of plasticity.[80] More significant evidence appeared in the 1960s and thereafter.

Brain reorientation is also well demonstrated in studies involving stroke patients who, although paralysed for years, were able to recover through the use of brain stimulating exercises. Michael Merzenich (*b.*1942) is an American neuroscientist who has been one of the pioneers of neuroplasticity for over three decades. He has made some of the most ambitious claims for the field, most notably that brain exercises may be as useful as drugs to treat diseases as severe as schizophrenia and that plasticity exists from cradle to grave.[81, 82]

A number of studies have linked meditation practices to differences in cortical thickness or density of grey matter. Richard Davidson (*b.*1951), an American neuroscientist at the University of Wisconsin, has led experiments in co-operation with the Dalai Lama on the effects of meditation on the brain.[83] His results suggest that the long—or even short—term practice of meditation results in different levels of activity in brain regions associated with such qualities as attention, anxiety, depression, fear, anger, the ability of the body to heal itself. These functional changes may be caused by changes in the physical structure of the brain.

To summarise:

1. The brain is plastic and not 'hard-wired' from birth.
2. therefore, by implication, the same cognitive habits (thoughts) will necessarily involve the same brain pathways and brain regions.
3. however, if the brain pathways and regions are damaged, then the same cognitive habits performed, under motivation, will 'carve out' new pathways involving different brain regions.
4. and critically, as regards a new continent of thought, new cognitive habits and new ways of thinking will also carve out new brain pathways of scientific materialism towards a more spiritual and universal worldview.

If we do not have an appropriate mental category or 'label' for something (such as a new scientific theory or philosophical doctrine) we are unlikely to perceive it and 'store' it in memory. This explains why theologians have, in general, opposed any new knowledge (especially scientific) that cannot readily be fitted into a category hardened by the self-same theological concepts and dogmas over innumerable years.

Readers should never form the impression that the suggestion is that the left hemisphere of the brain is being equated with the 'materialistic brain' and is by default inferior to the holistic thinking associated with the right hemisphere. Each hemisphere is of equal importance and value within the limits of its functioning and *within the context of its application.*

We have seen how ordinary intellectual activity moves along well beaten paths in the brain, but a new kind of mental effort calls for something very different. Just as a sportsman develops specific muscles for the task in hand, the esotericist must develop his 'mental muscles' to acquire a higher faculty of thinking. In both cases, the endeavour is fatiguing, painful, and discouraging.

The conventional analytical approach provides great detail about individual 'splinters' of truth at the expense of an understanding of the whole. In mainstream science this tends towards a materialistic and reductionist approach. In contrast, the holistic approach (the 'marriage' of left-brain with right-brain) forever maintains the organic unity and appreciates that the whole is always greater than the sum of its parts—albeit, with some loss of detail.

* * *

This chapter has outlined the paradigm of science constituting the basis for its dictates on mind and consciousness. The glorious vistas that are revealed when science broadens its outlook so as to transcend (not reject) its materialistic stance are then described. This forms the basis of the next chapter on the Mystery Teachings that resolve several conundrums that currently beset mainstream science.

NOTES

1. Louis de Broglie, 'The Role of the Engineer in the Age of Science', in *New Perspectives in Physics,* trans. A. J. Pomerans (New York: Basic Books, 1962), 213.
2. David Brewster, *A Short Scheme of the True Religion. Manuscript quoted in Memoirs of the Life, Writings and Discoveries of Sir Isaac Newton* (Edinburgh, 1850); See also, 'A short Schem [*sic*] of the True Religion', *The Newton Project,* Keynes MS 7, King's College, Cambridge, UK (published online, February 2002) <http://www.newtonproject.sussex.ac.uk/view/texts/normalized/ THEM00007> accessed 1 January 2020.
3. G. W. Leibniz, 'Clarification of the Difficulties Which Mr. Bayle Has Found in the New System of the Union of Soul and Body', in L. E. Loemker (ed.), *Philosophical Papers and Letters: The New Synthese Historical Library,* ii (Springer, Dordrecht, 1989). Also cited in C. I. Gerhardt (ed.), *Die Philosophischen Schriften von Gottfried Wilhelm Leibniz,* iv, 523–4.
4. E. Lester Smith, *Inner Adventures* (Wheaton, IL, The Theosophical Publishing House, 1988), 218.
5. David Skrbina, *The Metaphysics of Technology* (UK: Ashgate Publishing, 2015). An excellent review of this book is by David Lorimer, 'Technology and the Future', *Journal of the Scientific and Medical Network,* 122 (2016), 38–9.
6. Carl Vogt, Physiologische Briefe fuřGebildete aller Stände [Physiological Letters for Educated People of all Classes] (4th edn, rev. and enl, Gießen, Germany: J. Ricker, 1874), 354.

7. Francis Crick, *The Astonishing Hypothesis: The scientific search for the soul* (New York: Charles Scribner's Sons, 1994), 3.
8. Gerald Edelman, *Bright Air, Brilliant Fire: On the Matter of the Mind – A Nobel laureate's revolutionary vision of how the mind originates in the brain* (New York: Basic Books, 1992), 166.
9. Richard Westfall, *Never At Rest: A biography of Isaac Newton* (Cambridge: Cambridge University Press, 1995).
10. Steven Novella, *Your Deceptive Mind: A scientific guide to critical thinking skills*, 2012; and Michael Shermer, *Skepticism 101: How to think like a scientist*, 2013. Two CD courses with accompanying manuals from 'The Great Courses' <www.thegreatcourses.co.uk> accessed 30 November 2019.
11. Refer to M. A. Thalbourne, *A Glossary of Terms Used in Parapsychology* (London: William Heinemann, 1982).
12. Daryl J. Bem and Charles Honorton, 'Does Psi Exist? Replicable Evidence for an Anomalous Process of Information Transfer', *Psychological Bulletin*, 115/1 (1994), 4 –18.
13. W. Braud, *Distant Mental Influence* (Charlottesville, Virginia: Hampton Roads Publishing, 2003).
14. R. Sheldrake and P. Smart, 'A Dog that Seems to Know When Its Owner is Coming Home: Videotaped experiments and observations', *Journal of Scientific Exploration*, 14 (2000), 233–55.
15. L. L. Vasiliev, *Experiments in Mental Suggestion* (Studies in Consciousness) (Charlottesville, Virginia: Hampton Roads Publishing, 2002).
16. J. Utts, 'An Assessment of the Evidence for Psychic Functioning', *Journal of Scientific Exploration*, 10/1 (1996), 3–30.
17. Dean Radin and Diane Ferrari, 'Effects of Consciousness on the Fall of Dice: A Meta-analysis', *Journal of Scientific Exploration*, 5/1 (1991), 61–83.

18. Dean Radin, *The Conscious Universe: The scientific truth of psychic phenomena* (New York: HarperOne, 2009).
19. Society for Psychical Research <https://www.spr.ac.uk> accessed 24 December 2019. Impressive literature also exists in the Scottish Society for Psychical Research <https://www.spr.ac.uk/link/scottish-society-psychical-research> accessed 24 December 2019; and the American Society for Psychical Research <http://www.aspr.com> accessed 24 December 2019.
20. Dr Eben Alexander, *Proof of Heaven: A neurosurgeon's journey into the afterlife* (Great Britain: Piatkus, US: Simon & Schuster, 2012), 10.
21. Rajiv Parti (with Paul Perry), *Dying to Wake Up: A doctor's voyage into the afterlife and the wisdom he brought back* (UK: Hay House 2016); reviewed by David Lorimer, 'Deep Transformation', *Journal of the Scientific and Medical Network*, 122 (2016), 52–3.
22. Josh Weisberg, *Consciousness: Key problems in philosophy* (Cambridge, UK: Polity Press, 2014).
23. B. C. Goodwin, 'A Cognitive View of Biological Process', *J. Soc. Biol. Structures*, 1 (1978), 117–25.
24. Ulisse di Corpo and Antonella Vannini, *Syntropy: The spirit of love* (ICRL Press, 2015).
25. Representative examples over the past half-century are: C. E. M. Hansel, *ESP: A scientific evaluation* (New York: Charles Scribner's Sons, 1966), rev. edn, *ESP and Parapsychology: A critical re-evaluation* (Buffalo, New York: Prometheus Books), 1980; W. G. Roll and J. Beloff (eds), *Research in Parapsychology* (Metuchen, NJ: Scarecrow Press, 1980); Arthur S. Reber and James E. Alcock, 'Searching for the Impossible: Parapsychology's elusive quest', *American Psychologist* (13 June 019) <http://dx.doi.org/10.1037/amp0000486> accessed 18 November 2019.

26. Peter C. Gøtzsche, *Deadly Medicines and Organised Crime: How big pharma has corrupted health-care* (CRC Press, Boca Raton, Florida: Taylor & Francis Group, 2017); and Ben Goldacre, *Bad Pharma: how medicine is broken, and how we can fix it* (London: Fourth Estate, 2012). Both books were reviewed by David Lorimer, *Journal of the Scientific and Medical Network*, 114 (2014).
27. Joe Martino, 'Banned TED Talk: Rupert Sheldrake – The Science Delusion', Collective Evolution (10 February 2017) <https://www.collective-evolution.com/2017/02/10/banned- ted-talk-the- science-delusion-by-rupert-sheldrake-10-dogmas-that-exist-within-mainstream-science/> accessed 31 July 2020.
28. Rupert Sheldrake, 'TED "Bans" the Science Delusion' https://www.sheldrake.org/reactions/tedx-whitechapel-the-banned-talk accessed 31 July 2020. The complete talk can be accessed from the URL cited.
29. *The Scientific and Medical Network* <https://scimednet.org> accessed 17 November 2019.
30. Edwin Arthur Burtt, *The Metaphysical Foundations of Modern Physical Science* (Doubleday & Company, 1955).
31. 'Albert Einstein', Goodreads <http://www.goodreads.com/author/quotes/9810> accessed 14 November 2019.
32. This is a summarized account of a private communication to the writer by Professor Ellison in August 1996. An inaccurate version of the story is given in Kit Pedler, *Mind Over Matter* (UK: HarperCollins Distribution Services), 1982.
33. Rupert Sheldrake, *Morphic Resonance* (US: Park Street Press, 2009).
34. ———*The Presence of the Past: Morphic resonance and the habits of nature* (London: Fontana, 1989).
35. Wilder Penfield and Lamar Roberts, *Speech and Brain Mechanisms* (Princeton, New Jersey: Princeton University Press, 1959).

36. Rupert Sheldrake, *Morphic Resonance*, 162, 166, 173, 193–4.
37. 'Remembering and Forgetting', Rupert Sheldrake, *The Presence of the Past*, 197–209.
38. Myles Gough, 'Aboriginal Legends Reveal Ancient Secrets to Science', BBC News, 19 May 2015 <https://www.bbc.co.uk/news/world-australia-32701311> accessed 26 December 2019.
39. 'Aboriginal Dreamtime', Artlandish Aboriginal Art Gallery <https://www.aboriginal-art- australia.com/aboriginal-art-library/aboriginal-dreamtime> accessed 26 December 2019.
40. Rupert Sheldrake, *Dogs That Know When Their Owners are Coming Home, and Other Unexplained Powers of Animals* (London: Hutchinson), 1999.
41. Morris Kline, *Mathematical Thought from Ancient to Modern Times* (1972; New York: Oxford University Press, 1990), i, 18–21.
42. David Brewster, *The Life of Sir Isaac Newton* (New York: Harper & Brothers, 1831), 198 [online]<https://books.google.co.uk/books?id=Jq43zQEACAAJ&printsec=frontcover&source=gbs_ge_ summary_r&cad=0#v=onepage&q=verdict&f=false> accessed 10 July 2020. See also Richard S. Westfall, Never at Rest: A biography of Isaac Newton (Cambridge University Press, 1980), 698 et seq.
43. Gustaf Strömberg, *The Soul of the Universe* (1940; 2nd edn, Philadelphia: David McKay Company, 1948).
44. ——*op. cit.*, vii-viii.
45. Rollin McCraty, *Science of the Heart: Exploring the role of the heart in human performance*, 2 vols (HeartMath Institute) i 2001; ii 2015.
46. Bruce Goldstein, 'Cognitive Psychology', Wadsworth, Cengage Learning (2011), 374.
47. Frederic P. Miller, Agnes F. Vandome, and John McBrewster, *Confirmation Bias* (Mauritius: VDM Publishing, 2009), 1.

48. What follows is a summary from Edi D. Bilimoria, *The Snake and the Rope: Problems in Western science resolved by occult science* (Adyar, Chennai: Theosophical Publishing House), 276–7.
49. Graham Coleman and Thupten Jinpa (eds), *The Tibetan Book of the Dead:* First complete translation, with commentary by the Dalai Lama, trans. Gyurme Dorje (New York: Penguin Classics, Deluxe edition, 2007).
50. Kevin Williams, 'The Tibetan Book of the Dead and NDEs' <https://www.near-death.com/ religion/buddhism/ tibetan-book-of-the-dead.html> accessed 14 November 2019.
51. Sir E. A. Wallis Budge (Egyptian text transliteration and trans.), *Egyptian Book of the Dead* (*The Papyrus of Ani*) (New York: Dover Publications, 1967).
52. Plato, *The Republic,* trans. with introduction by R. E. Allen (New Haven: Yale University Press, 2006).
53. Robert A. Monroe, *Journeys Out of the Body* (UK: Souvenir Press, 1989).
54. Arthur J. Ellison, *The Reality of the Paranormal* (London: Harrap, 1988), 75–8, 79, 81, 83.
55. S. J. Muldoon and H. Carrington, *The Projection of the Astral Body* (London: Rider, 1929).
56. See E. Lester Smith, *Inner Adventures* (Wheaton, Illinois: Theosophical Publishing House, 1988), 132– 55 passim.
57. Lecture by Professor Dr Peter A. E. Stewart, DSc, FREng, FRAeS, FInstP, 'An Engineering Life' (9 January 2008) <http://rpec.co.uk/rpec_new/pages/_Exc-11.html> accessed 14 November 2019.
58. Ingo Swan, 'The Ingo Swann 1973 Remote Viewing Probe of the Planet Jupiter' <https://www.thelivingmoon.com/44cosmic_wisdom/01documents/Ingo_Swann_Remote_Viewing_Jupiter.htm> accessed 14 November 2019.

59. 'STAR GATE [Controlled Remote Viewing]' <https://fas.org/irp/program/collect/stargate.htm> accessed 14 November 2019.
60. Jessica Utts, 'Appreciating Statistics', *Journal of the American Statistical Association*, 111/516 (2016), 1373–80.
61. Annie Besant and C. W. Leadbeater, *Occult Chemistry: Investigations by clairvoyant magnification into the structure of the atoms of the periodic table and of some compounds*, ed. C. Jinarâjadâsa, assisted by Elizabeth W. Preston (3rd edn, Adyar, Madras: Theosophical Publishing House), 1994.
62. Stephen M. Phillips, *The Mathematical Connection Between Science and Religion* (UK: Anthony Rowe Publishing, 2009). This book of peerless calibre was reviewed by Edi Bilimoria, *Journal of the Scientific and Medical Network*, 122 (2016), 52–3.
63. Sylvia Cranston (compiled and ed.), *Reincarnation: The phoenix fire mystery* (Pasadena, California: Theosophical University Press, 1994). The foreword is by Elisabeth Kübler-Ross MD (1926–2004), the Swiss–American psychiatrist and pioneer in near-death studies.
64. Ian Stevenson, *Reincarnation and Biology: A contribution to the etiology of birthmarks and birth defects* – Vol. 1: Birthmarks; Vol. 2: Birth defects and other anomalies (Westport, Connecticut: Praeger Publishers, 1997). For a short, non-technical version, see Ian Stevenson, *Where Reincarnation and Biology Intersect* (Westport, Connecticut: Praeger Publishers, 1997).
65. ——*Cases of the Reincarnation Type* – Vol. I: *Ten cases in India*, 1975; Vol. II: *Ten cases in Sri Lanka*, 1978; Vol. III: Twelve cases in Lebanon and Turkey, 1980; Vol. IV: *Twelve cases in Thailand and Burma* (Charlottesville, US: University of Virginia Press), 1983.
66. One of the best detailed accounts is John G. Fuller, *The Ghost of Flight 401* (London: Corgi Books, 1979). An excellent summary is by E. Lester Smith, *Our Last Adventure*

(London: Theosophical Publishing House, 1985), 40–6. Lester Smith was a Fellow of the Royal Society.

67. C. G. Jung, *The Archetypes and the Collective Unconscious* (UK: Taylor & Francis, 1991).
68. There are numerous references to this. Just two examples explaining the role and interaction of planets, comets, the solar system, and the Milky Way are in H. P. Blavatsky: *The Secret Doctrine* (The Theosophical Publishing House, First Quest Edition, 1993), Vol. I, 'The Modern Nebular Theory', 593-4; and *The Collected Writings* (The Theosophical Publishing House, Wheaton, Illinois, 1974), 'Transactions of the Blavatsky Lodge', 402.
69. J. S. Gordon, *The Rise and Fall of Atlantis and the Mysterious Origins of Human Civilization* (London: Watkins Publishing, 2008).
70. Jessica Utts, 'Appreciating Statistics', *Journal of the American Statistical Association*, 111/516 (2016), 1373–80.
71. Nikola Tesler <https://www.amazon.com/non-physical-phenomena-progress-centuries-exis-tence/dp/B01M3S41OM> accessed 15 November 2019.
72. David Lorimer, 'The Galileo Commission: Towards a post-materialist science. An invitation to look through the telescope', *Journal for the Study of Spirituality*, 9/1 (2019).
73. Charles T. Tart (ed.), *Altered States of Consciousness* (New York: Doubleday, 1972).
74. David J. Chalmers, 'Facing Up to the Problem of Consciousness', *Journal of Consciousness Studies*, 2/3 (1995), 200–19; see also <http://consc.net/papers/facing.html> accessed 20 November 2019.
75. ——*The Conscious Mind: In search of a fundamental theory* (New York and Oxford: Oxford University Press, 1997).
76. Ervin Lászl6, *Science and the Akashic Field: An integral theory of everything* (Rochester, Vermont: Inner Traditions, 2007), 76.

77. — —*op. cit.*, exordium, xii.
78. A. Pascual-Leone, et al., 'Characterizing Brain Cortical Plasticity and Network Dynamics Across the Age-Span in Health and Disease with TMS-EEG and TMS-f MRI', Brain Topography, 24 (2011), 302–15.
79. William James, *The Principles of Psychology* (US: Dover Publications, 2000).
80. K. S. Lashley, 'Basic Neural Mechanisms in Behavior', Address of the President of the American Psychological Association before the Ninth International Congress of Psychology at New Haven, 4 September 1929. First published in Psychological Review, 37 (1930), 1–24.
81. Dean V. Buonomano and Michael M. Merzenich, 'Cortical Plasticity: From synapses to maps', *Annual Review of Neuroscience*, 21 (1998), 149–86.
82. M. M. Merzenich, R. J. Nelson, M. P. Stryker, M. S. Cynader, A. Schoppmann, and J. M. Zook, 'Somatosensory Cortical Map Changes Following Digit Amputation in Adult Monkeys', *Journal of Comparative Neurology*, 224/4 (1984), 591–605.
83. Richard Davidson and Antoine Lutz, 'Buddha's Brain: Neuroplasticity and meditation', IEEE Signal Processing Magazine (January 2008).

Part Two

The Mystery Teachings of All Ages

Mystery traditions lie at the roots of both Western and Eastern cultures, taught by the ancient masters of wisdom who laid the foundations for the world we now live in.

'Mystery Teachings' is the generic term given to esoteric and sacred teachings concealed within the rituals, allegories, and observances of all ages, mostly kept secret from the profane or uninitiated masses and concerning the origin of things, the nature of the human spirit, its relation to the body and method of its purification and restoration to a higher life. These secret teachings are also known as theosophy, esotericism, occultism, and the esoteric doctrine.

These teachings about the invisible laws of nature, and the powers latent in man, have been taught in the Mystery Schools of great civilizations (like India, Egypt, Persia, Greece, and Mexico) since the dawn of time. Because the teachings concern the inner workings and secrets of nature and man, rather than the outward form, the information is revealed only to students who are suitably qualified in terms of nobility of character, moral attainment, and intellectual aptitude, so that they would not be misused for selfish or personal reasons.

The Mysteries in every country are divided into the 'Lesser' and the 'Greater'. In Egypt, the Lesser Mysteries were sacred to Isis and the Greater Mysteries to Serapis and Osiris. In Greece, the Lesser Mysteries were the Eleusinian and the Greater Mysteries included the Eleusinian and the more advanced Orphic and Dionysian. In all cases, the Lesser Mysteries largely comprised dramatic rites or ceremonies alongside some teaching. The Greater Mysteries were composed of study and the doctrines taught in them were proved later by personal experience in initiation. The first philosophers of antiquity were all initiated in the Mystery Schools.

The mysteries of cosmogony and nature were personified by the priests and neophytes, who enacted the parts of various gods and goddesses, repeating scenes from their respective lives. Their hidden meanings were then explained to the candidates for initiation and incorporated into philosophical doctrines.

Mystery traditions lie at the roots of both Western and Eastern cultures, taught by the ancient masters of wisdom who laid the foundations for the world we now live in, first to chosen disciples who then disseminated the teachings on a wider front. It has become common in the West to turn to Eastern sources for real understanding and enlightenment; however, as constantly emphasised in this book, Western culture, albeit highly materialistic, is underpinned by a tradition of incredible wisdom and significance which we need now more urgently than ever to rediscover and reclaim.

The Canadian occultist, philosopher, and prodigious researcher of esoteric and mythological lore, Manly P. Hall remarked in 1988 that, 'we are now coming to the end of the twentieth century, and the great materialistic progress which we have venerated for so long is on the verge of bankruptcy.' Therefore, 'to avoid a future of war, crime, and bankruptcy, the individual must begin to plan his own destiny, and the best source of the necessary information comes down to us through the writings of the ancients.'[1]

Exhaustive research and historical scholarship have shown how esotericism lies at the core of science and spirituality, characterising and fecundating both Eastern and Western thought from antiquity to the present age.

Definitions – Esoteric and Exoteric

Spirit connotes the eternal and immortal element in all of nature, man and the universe. In man, this refers to some kind of surviving entity after the death of the mortal, physical body.

A *philosopher* is one who takes full cognisance of the spiritual dimension of life and the universe. The term derives from the Greek *philo* 'loving' and *sophia* 'wisdom'.

'*Theosophy*' comes from the Greek *theosophia* meaning 'Wisdom-Religion' or 'Divine Wisdom' derived from the compound *theos*- 'a divine being', and *sophia*. Its doctrines are not a religion but divine consciousness or divine science and, in practical terms, divine ethics.

Mysticism, common to all religions and cultures, represents a state of transformation resulting in union with a higher power known by terms such the Absolute, the Infinite or God. A *mystic* is one who experiences intimations or intuitions of the existence of inner, invisible and superior worlds (transcending the physical) and who attempts self-conscious communion with them and the beings who inhabit these worlds.

Esotericism is a generalising term for the entire body of the esoteric philosophy or doctrine, derived from the Greek *esôterikos* (esotericos) 'inner', 'concealed', 'hidden', 'secret', 'pertaining to the innermost'. Its opposite is *exoteric* or profane.

The *esoteric doctrine* is the body of mystical and sacred teachings—about the universe, nature and man, reserved for students of high and worthy character. Since time immemorial, it has been held in the guardianship of the seers, sages, initiates, and adepts who periodically disseminate portions of it to the world when the intellectual and spiritual needs of mankind are ripe. If all the great religions and philosophies are critically examined, the fundamental principles in each will be found to be identical. All these world-religions and world-philosophies contain the entirety of the esoteric doctrine, but it is invariably expressed in exoteric form.

When taken to extremes, the exoteric, literal doctrinaire statement of sacred esoteric truths becomes a source of utter confusion, distortion, mayhem and, not to say, hatred begetting violence. One need only think of the innumerable wars that have been fought *in the name* of religion 'because it says so in the Bible' or because 'the Qur'an says it'. Such dogmatic, dead-letter interpretations simply display an utter failure to see the deeply allegorical and symbolic nature of the scriptures that must be understood.

One also needs to be wary of reading too much into the texts that have been edited and doctored over the ages.

Blind belief and worship of the outer form are critical to maintaining the status quo of the exoteric form of religions by religious authorities. There is no shortage of such superstitious genuflection by the masses in front of statues and portraits, as in India or in predominantly Catholic countries. All this clearly shows why the great instructors of all religions, especially in ancient times, rarely wrote down their own teachings and on the few occasions when they did, it was in highly cryptic allegorical form that could be understood only by those acquainted with the inner meaning. More often their instruction was transmitted orally to chosen disciples to disseminate in the manner best suited to their times and culture. They knew that once something was written down, it would be 'set in stone', the inner living message suffocated by the outer crust of rigid interpretation.

Whereas esotericism *reveals* the inner meaning and truth, exotericism masks or *'re-veils'* the truth by presenting it in the form of the outward and popular (literal) interpretations found in the public rituals and ceremonies of orthodox religions. Esotericism is also known as the 'Doctrine of the Heart', namely, seeing into the inner meaning of a subject or object using the awakened spiritual eye while exotericism is referred to as the 'Doctrine of the Eye', that is, seeing the outward form.

Occultism takes esotericism a stage further by dealing, not just with the characteristic nature of all the energies and forces in the universe, both objective and subjective, but also the techniques by which they work, are evoked, and can be controlled. We need to define, and then justify, 'occultism' with utmost care because few words occasion more mindless hysteria from people in both the scientific and religious orthodoxy. One feels rather hesitant about using a word that has unpleasant and wrongful associations with dark practices. All the meaningless and asinine 'knee-jerk' associations with black magic and witchcraft notwithstanding, the word 'occult' simply means 'secret', 'hidden', or 'concealed'. It derives from the Latin *occultus* meaning to 'hide' or 'cover over'.

Occult science is a generic term referring to the Hermetic or esoteric sciences, which explore the essential, hidden secrets of nature—physical and psychic, mental and spiritual—rather than just her outward appearance. Occultism is therefore a generalising term for the entire body of occult sciences.

In essence then, mysticism is concerned primarily with feeling, esotericism with meaning and understanding, and occultism with function. *Esotericism is the intellectual arm of mysticism and occultism is the practical, or functional arm of esotericism.*

Mankind has never ceased asking profound questions about whether there was ever a time that the universe was not, how it came into existence and his own place in it. Philosophy, religion, or science can be studied either from the conventional or orthodox perspective of intellectual speculations based on physical appearances or from the occult standpoint, focussing on unearthing and actively harnessing the inner, core principles buried in the myriad forms in which such subjects are presented.

The misuse of occultism in the hands of unscrupulous practitioners has no bearing on its rightful practice by philanthropic persons motivated by a selfless desire for universal good.

Among the ancient civilizations resplendent during their acme but then degenerating through misuse of occult knowledge, we can especially cite the prehistoric Atlanteans, or the Egyptians and Native Americans. The iniquity of these civilisations during the final stages of their epoch stands in inverse proportion to their majesty during their zenith. In the case of the Egyptians and Native American Indians, white magic degenerated into sorcery and subsequent black magic with attendant appalling rituals resulting in the eventual degeneration of those once lofty civilisations. (See below for the difference between white magic and black magic.) The same applies with even greater force to the rise and fall of the Atlantean civilization and the catastrophic destruction over long periods of time of that once majestic continent whose existence is forever an uncomfortable enigma to the 'fact finding' archaeologist and archival historian.[a]

In modern times, the blatant abuse of occult knowledge by the leaders of the Third Reich should provide ample evidence of the appalling catastrophes, and the tragic consequences, to the evil-doers resulting from attempting to twist nature's secrets for selfish, power-seeking motives.[2] There is no better example of their misunderstanding and criminal misuse of occult knowledge than the Nazis' reversal of that ancient and sacred symbol, the Swastika.

White Magic, Sorcery, and Black Magic – the Crucial Differences

Terms such as 'white magic' and 'black magic' are apt to cause needless misunderstanding. Quite simply, white magic is referred to in occult literature as choosing to follow the Right-Hand Path. It refers to a totally altruistic, philanthropic, and utterly impersonal use of occult powers directed solely towards the moral and spiritual elevation of humanity.

---

a The occult literature provides authentic and referenced accounts of the prehistoric Atlantean civilization and the destruction of the associated continent. One of the best modern books is by John Gordon, *The Rise and Fall of Atlantis and the Mysterious Origins of Human Civilization* (Watkins Publishing, 2008).

On the other hand, the acquisition of power and control over others in unclean practices, such as witchcraft and voodoo, for personal gain is known in general as sorcery and referred to as following the Left-Hand Path. At its worst, it is black magic and will exact a terrible price of *self-generated* suffering upon the wrongdoer.

It is of the utmost importance to understand that highly evil people can be decidedly evolved except that they have chosen to take the Left-Hand Path of black magic instead of the Right-Hand Path of white magic. Although we tend to regard evolved persons as spiritually beneficent individuals and a major force for the good of society, yet the evolutionary status of individuals says nothing about their *spiritual* status.

What Are Psychic Powers?

'Psychic' and 'psychic powers' are also terms encountered widely in arcane literature. Psychic, from the Greek *psyche*, refers to the intermediate nature of man—that which lies between spirit and body, namely his soul in the generic sense. They correspond to the lower strata of man's soul and are the lower manifestation of occult powers. One cannot consciously exercise psychic powers without some degree of esoteric knowledge. In appropriate circumstances, with guidance and training, such powers can be evoked, but the dangers to health and sanity are immense if any attempt is made to force their development prematurely. When misused for selfish reasons, to gain power and control over people, the result is sorcery, or black magic.

The Language of the Mysteries and the Use of 'Blinds'

The Mystery language was the sacerdotal secret jargon employed by initiated priests and used only when discussing sacred things. Every nation had its own 'Mystery tongue' unknown save to those admitted to the Mysteries. In consequence, those who were versed in esoteric and occult matters would be able to unwrap the outer covering and find the terminology to be terse and clear—revealing the kernel of truth within. But for the profane, the terminology would seem meaningless gobbledygook.

Thus, a frequent charge about the language and literature of the Mysteries (chiefly, occultism) is that they are full of waffle and jargon unintelligible to the ordinary person. However, is it not the case that each discipline draws upon a language best suited to communicate its message? Of course, merely using jargon in no way implies that the user understands its meaning (and in fact jargon used in that way is waffle and serves only to confuse).

Moreover, such 'jargon' fulfils a double necessity: to communicate facts to the worthy, and protect those facts from falling into unscrupulous hands and being perverted. For this reason, the Mystery Teachings were conveyed in an outer wrapping in what is known as 'blinds': a word, or symbol, used as an allegorical surrogate or proxy for its real meaning.

It bears emphasising that occult facts are highly dangerous in unworthy hands. So, the *apparently* gibberish nature of occult terminology is simply a way to ensure that the secrets about the inner workings of nature, and latent powers in man, will not fall into undeserving hands but will be understood only by those who are worthy to receive them.

In Hall's words: 'In a single figure, a symbol may both reveal and conceal, for to the wise the subject of the symbol is obvious, while to the ignorant the figure remains inscrutable. With the needs of posterity foremost in mind, the sages of old went to inconceivable extremes to make certain that their knowledge would be preserved. [For example] their knowledge of chemistry and mathematics they hid within mythologies which the ignorant would perpetuate, or in the [geometrical proportions of] spans and arches of their temples which time has not entirely obliterated.'[3]

It is heart-breaking, nowadays, to witness the destruction of the ancient temples and archaeological sites in Iraq and Syria (like the ancient Semitic city of Palmyra) by the murderously fanatical 'Islamic State'—having naught to do with the true teachings of the Prophet of Islam. One therefore shudders to think what would happen if the ancient secrets about nature's finer forces and powers were to fall into their unworthy hands like totalitarian dictators. The need for secrecy is apparent.

The Charge of Secrecy

Next, we deal with the related charge of secrecy: that besides being expounded through allegories and blinds, Mystery Teachings were never publicly stated. Even what was stated had vital elements missing and the seeker had to go to extraordinary lengths to find the hidden secrets of the teachings.

The true occultist contributes immensely to the sum total of human knowledge and wisdom, whilst maintaining silence about his personality and the organisation or brotherhood to which he belongs. What is the nature of that silence?

It is symbolised in the Mysteries by the closing of the eyes and the closing of the mouth. Closing of the eyes has nothing to do with inducing temporary blindness. Quite the contrary: by shutting out external distractions, the internal vision becomes magnified and clarified—every serious meditator knows and experiences this. As for the closing of the mouth, this is not just because the secrets may not be communicated but that they *cannot* be communicated. It is as impossible to describe in words the taste of wine to someone who has never drunk it, as it is to impart secrets to a neophyte who has not been initiated into their meaning.

As for the complaint that a vital piece of the esoteric jigsaw is being left out of the teachings or the missing clue deeply hidden in blinds and symbols, this is because the initiated philosophers realised that once man understands the complete workings of any system, he may accomplish a prescribed end (often motivated by personal power or greed), but he would not necessarily be qualified to deal with, or control, the effects so produced.

In Chapter XI of the Rosicrucian manifesto, *Confessio Fraternitatis,* we find this warning—a fitting example of the great misuse of power by the obliterating forces of dogmatic Christian ideology:

When to a man is given power to heal disease, to overcome poverty, and to reach a position of worldly dignity, that man is beset by numerous temptations and unless he possess true knowledge and full understanding he will become a terrible menace to mankind.

The alchemist who attains to the art of transmuting base metals can do all manner of evil unless his understanding be as great as his self-created wealth. We therefore affirm that man must first gain knowledge, virtue, and understanding; then all other things may be added unto him. We accuse the Christian Church of the great sin of possessing power and using it unwisely; therefore we prophesy that it shall fall by the weight of its own iniquities and its crown shall be brought to naught.[4]

It would not be entirely unreasonable to argue that such prophecy has indeed come to pass given the moribund state of orthodox Christianity in the public eye these days. These sequestered schools or societies were typically the Alchemists, Rosicrucians, Freemasons, Cathars, Gnostics, Templars, and Theosophists. Each of these movements was, at one time or another, virulently opposed by the Catholic Church, which thought nothing of persecuting, torturing, burning, butchering or, at best, excommunicating its members. Even the modern Theosophical Society received its fair share of malicious attacks from the Christian missionaries in nineteenth-century India. It is for this reason that such schools, particularly during the Middle Ages, had to pursue their spiritual and esoteric functions under the cloak of disguises and blinds.

In contrast, other major religions, especially in the East, have shown far more tolerance towards their esoteric brothers than official Christianity. But even in Christianity we easily discern the use of blinds, otherwise known as sacraments, which are religious ceremonies regarded as an outward and visible sign of an inward and invisible reality, like baptism, confirmation, the Eucharist, etc.

Even a cursory glance at the current world situation, where we see a rampant craze for power and its misuse in the hands of unscrupulous politicians, business leaders and pseudo spiritual cult figures—should make it crystal clear why long periods of arduous probation were imposed on the occult neophyte—so that the knowledge of how to become *as the gods* might remain the sole possession of the worthy.

It is obvious that knowledge is power and those who would fathom nature's secret processes or solicit the company of the adepts, would have the power to wreak untold havoc in the world, unless they possessed full control and absolute power *over themselves*.

It is worth pointing out that occult mysteries have been protected almost as effectively, if entirely unwittingly, by modern scientific scepticism as they have been in the past by blinds, sacraments, and secrecy.

Who Are Adepts and Initiates?

An initiate is one to whom the Mystery Teachings have deservedly been imparted at a level appropriate to his grade of development. An adept is one who has attained full mastery over himself and the Mysteries (i.e., secrets of nature) to which he has had access and been entrusted. It stands to reason, therefore, that an adept is a fully fledged occultist.

The term 'initiate' derives from the Latin *initiatus* meaning 'to begin' or 'originate'. It is the designation of anyone who was received into, and had revealed to him, the mysteries and secrets of the occult sciences, whether this be through traditions of the East or the West—the Qabbalah, Rosicrucianism or Masonry being prime examples of the latter. Traditionally, the arcane knowledge was imparted to the chosen neophyte by the hierophants ('One who explains sacred things') of the Mysteries. In modern times there are still teachers capable of instructing initiants worthy of the teaching, but these are necessarily few and far between. Note the distinction between the 'initiant', one who is beginning, or preparing for, an initiation and an 'initiate', one who has successfully passed at least one initiation.

An adept is one who has obtained the highest initiation. The term derives from the Latin *adeptus* meaning 'having reached' or 'attained'. It signifies one who has reached the stage of a master in esoteric wisdom and occult philosophy. In everyday parlance, we refer to anyone who is highly skilled, and attained mastery over their profession, as being adept at it. There is no logical reason why such adeptship cannot include not just mastery over a chosen skill, but full mastery over the complete human nature at all levels, physical, emotional, psychic, and mental.

The adept may promulgate as much of such wisdom as he is lawfully entitled to do, either on his own or by admission into a fraternal organisation as one of its principal representatives, or through chosen initiates as his messengers and envoys such as H. P. Blavatsky.

Adeptship represents the highest attainment of the human kingdom after which the adept may progress to states and planes of consciousness out of the human kingdom into still higher kingdoms of nature. An equivalent term to adept is 'master' or the generic Indian term 'mahatma' (a specific august being), a Sanskrit term meaning 'great soul'.

That such mysterious men did exist there can be little doubt, as their presence is attested to by scores of reliable witnesses. There are indications that a certain degree of organisation existed among them. The most powerful of such hierarchies in the West and the Middle East were the Rosicrucians, the Illuminati, and certain Arabian and Syrian sects and, in the East, the Great White Brotherhood. They are still to be found by those who have qualified themselves to contact them.

The perennial wisdom tradition has always taught about intelligences ('gods' from planetary and solar systems higher than our own) watching over, and guiding, the spiritual evolution and consciousness of nascent humanity and intervening at critical moments to assist its evolutionary progress. This idea has captured the imaginations, and fired the aspirations, of seekers past and present, both in the East and West. However, this same idea has become prey to the most serious misunderstandings and misconceptions. There is no shortage of self-appointed 'masters' coveting personal aggrandisement along with power and control. This is why the whereabouts, and other details, of real masters are closely guarded secrets open only to those initiates and advanced seekers qualified to contact them. An exception to this general rule may be Madame Blavatsky, principal founder of the modern Theosophical movement. She was the first to reveal openly to the public some of the facts of the Masters and to confirm that the movement she spearheaded, and its teachings, were the project of her two principal instructors. Alas, this excessive divulging of facts, which Blavatsky herself regretfully admitted should never have been made public, has caused seismic disturbances within the Theosophical Society itself, not to mention the personal calumny, charges of fraud and character assassinations that she personally had to bear.

How Can Adepts Be Contacted?

The masters can be reached only by those who are capable of transcending the limitations of the material world. Purity of motive is the signal that will attract the attention of an adept, however weak or unqualified one may be in other respects. Reassuringly, it is a spiritual law that any advanced being must seek out a pure aspirant for wisdom. This is the real truth behind the popular saying: 'Take one step towards the divine and the divine will take nine steps towards you', which really means, 'God helps those who help themselves'.

How an advanced being seeks out and contacts the earnest aspirant is never at our personal convenience or behest. The contact is very rarely in person and mostly in the form of synchronous occurrences that impel the inner nature towards the higher life—for example, finding just the right book or acquiring an exceptional friendship at just the right time in life. Needless to say, any accelerated growth process invariably involves suffering and hardship in the personal life during the period of transition.

The Mystery Schools and Their Teachings

Let us outline the necessary qualifications and essential secrets divulged to chosen candidates in the Mystery Schools, whether in the Brahmanical Schools of India, the ancient Persian and Egyptian Schools, the Odinic Mysteries of Scandinavia, the Eleusinian, Orphic and Dionysiac Mysteries of Greece, or the Druidic Schools of ancient Britain.

The principal objective of the Mysteries is to enable the struggling human being to reawaken, and re-claim, the spiritual fire lying dormant within himself but suffocated by the lust, selfishness and decadence of his baser nature. The whole purpose of the secret processes involved was to enable certain centres of the brain to be stimulated so that the consciousness could be liberated and extended to invisible landscapes that remain entirely opaque to everyday consciousness.

The ancients have always taught that a man does not know the gods (divine forces and powers) by logic and intellect alone or by external activities and influences, *but primarily by realising the presence of the gods within himself.*

Never in their entire history have the Mysteries ever catered to the personal and emotional nature and needs of the candidate. It is precisely to free the soul of limitation, to purify the heart and discipline the mind, that the Mystery training is so severe, for in initiation only spiritual strength, underpinned by unimpeachable strength of character, can withstand the ordeal.

Location of the Mystery Schools – Ancient and Modern

Shamballa is the Sanskrit place-name of the mythical location and abode of the divine *Kumaras*, those mighty intelligences who are the real guides of our planetary life. Its abode is said to exist in invisibly ethereal substance somewhere to the north of the Tibetan plateau. Shamballa is best regarded as being primarily visionary and spiritual, and only secondarily, physical or geographical.

Shamballa is the mystic heart-centre of the Earth's spiritual body. From it the esoteric life-blood flows into organic Mystery-centres of esoteric and occult learning and instruction. Every Mystery-centre is an organic focus owing spiritual allegiance to the mystic heart-centre. The Mysteries were taught in specially created schools by priest— hierophants who developed extraordinary techniques to assist candidates achieve higher states of consciousness. However, a Mystery School is not necessarily dependent on location. Rather, it is an association or brotherhood of spiritually disciplined individuals bound by one common purpose—the spiritual and moral elevation of humanity born of love and wisdom.

Modern examples would, arguably, be the Theosophical Society, on the one hand, which has its international head-quarters in India and the Scientific and Medical Network which has no physical headquarters of its own.

It is a fact that certain locations appear to be more favourable to success in spiritual matters than others. Why were the ancient seats of the Mysteries almost invariably in rock-temples or subterranean caves, in forests or mountain passes, in pyramid chambers or temple crypts? Because the thought currents and atmosphere prevailing in such remote locations would be calmer, more peaceful and cleaner the farther removed from the 'maddening crowd' of cities. So rarely, if ever, would one find a seat of esoteric training near a large metropolis.

Do Mystery Schools, or Mystery-centres exist these days? There is no doubt that they have withdrawn from public knowledge. That does not automatically imply that the perennial link with Shamballa has been severed but the link is now preserved under a veil of secrecy. Such Mystery-centres are to be found the world over even though the exact location of these centres, guarded with zealous care by their protectors, is undiscoverable except by the worthy. Owing to the increased need for light and truth nowadays, 'the esoteric groups of Mystery-Schools are perhaps more numerous today than they have been for thousands of years.'[5]

Emissaries of the Mystery Schools and Their Fate

In the broadest sense, emissaries are those who have made their appearance on the world stage with the express purpose of revealing a portion of the Mystery teachings appropriate to the epoch in question in order to elevate the consciousness of humanity, whether this be in the realms of religion, philosophy, science or culture. Their presence has exerted a profoundly beneficent influence on human society and changed the course of history. (Admittedly, human society and history have also been radically altered by tyrants and dictators who have wielded unimaginable power. This again suggests some degree of contact with occult teachings. In such cases these teachings have been grotesquely perverted and misused for selfish gain.)

The central issue is really a matter of internal contact, by way of spiritual resonance between the Higher Self of an aspirant and the radiating consciousness of august beings unknown to the world at large. And there is no denying the radical and lasting change that has occurred in the course of science and music from such 'resonant contact' by great scientists and musicians.

How has the world at large treated its ambassadors of truth? Their lot has rarely been a happy one, particularly in the Middle East and the West. Many sages and saints were streets ahead of their times in mental, intellectual, and spiritual capacities. And so, with few exceptions, like the Buddha, they had to suffer the resentfulness and suspicions of the unthinking multitude, who labour under the delusion that advanced ideas and divine truths can be forever obliterated by murdering, persecuting, incarcerating, exiling or slandering those who sought to bequeath them to the world.

The iniquities perpetrated against the emissaries of truth makes depressing reading. Among the prophets and religious mystics we see Zarathuśtra killed by the assassin's spear; the crucifixion of the Christ and the flight from Mecca of Muhammad (*circa* 570–632). Then, among the great philosophers, Socrates (*circa* 469/470–399 BC) executed by drinking poison hemlock; the defenceless Hypatia (*circa* 350/370–415 AD) dragged from her chariot by a Christian mob; and Giordano Bruno (1548–1600), the Italian Dominican friar and cosmological theorist burned at the stake. In modern times the occultist H. P. Blavatsky and the philosopher and sage Peter Deunov (1864–1944) were persecuted; the great Indian political leader Mahatma Gandhi (1869–1948) was assassinated.

The list is a long and sad one. Even the legendary Pythagoras (*circa* 570 BC–*circa* 495 BC) was not spared. According to some, he was murdered with his disciples; others say that he was burned alive by his enemies. An ignorant world has invariably persecuted those who understood the secret workings of nature, seeking in every conceivable manner to exterminate the custodians of divine wisdom.

But why is such ingratitude shown towards humanity's instructors for which humanity has, and continues, to pay a terrible price? The outspokenness of geniuses evokes personal and political hostility. In every case it is because of the clash between the 'new' and the 'old', between the visionary and the dyed-in-the-wool establishment clergy, between the prophet of the new and the priest of the past. Quite simply, the emissaries of truth brought forth a living and new message that threatened the status quo of the *establishment*—the church, the synagogue, the mosque and the temple—entrenched in their own past.

What was the 'crime' committed by these advanced beings who strove to uplift the bar of humanity's threshold of consciousness? Each one of them encouraged people to substitute blind belief for a questioning attitude, not to accept the status quo as the final verdict on truth, to seek fresh insight and to break through the crust of outworn ideas, customs, and rituals. As the establishment sees this tidal wave of change, it girds up its loins in ever more ferocious and fanatical attempts to push back the surge, but truth is too powerful for it and although the fatalities may be great, truth will eventually prevail.

Preparation for Initiation

Preparation, or pre-qualification, for the initiation degrees into the Mystery Schools involved three disciplines: (*a*) self-purification; (*b*) building up of the intellectual powers; and (*c*) mastery over the animal nature.

Any system or mechanism, from the most basic to the most subtle, needs to incorporate the cathartic arts, that is, arts such as prayer, meditation, contemplation, and ritual to cleanse the disciple so he is ready to receive the Mysteries that shall be imparted to him.

The intellectual powers must be fully functioning in order to discriminate truth from falsehood, reality from appearance, dogma from certainty—and never more so in the present climate of façades and fake news.

Control and mastery over the animal nature is a *sine qua non* to ensure that the knowledge imparted be directed towards altruistic and philanthropic ends rather than the appeasement of the lower, sensual nature. Note that all three pre-qualifications must be developed to a high degree: over-development of any one does not compensate for deficiency in another.

Degrees of Initiation

Every country has its own methods of preserving the knowledge and tradition of the Mysteries. Using the cult of Mithraism as an example, the three degrees of initiation are now described.

In the first degree the candidate was given a golden crown on the point of a sword and instructed in the mysteries of the hidden power of Mithras. The sword probably symbolised the cutting edge of truth wielded by the power of the intellect.

In the second degree the candidate was given the armour of purity and intelligence (a potent combination) and banished into the darkness of subterranean pits to fight the beasts of passion, degeneracy, and lust.

In the third degree he was given a cape upon which were drawn the signs of the zodiac and other astronomical symbols. To those acquainted with their use, such strange hieroglyphic figures and symbols gave unlimited power and control over the unseen forces of nature.

Occult Science in Contrast with Natural Science

The scientific method moves from the particular to the general, rather like assembling individual pieces of a jigsaw puzzle. Discrete observational data are collected and gradually fitted into a general picture. This is the 'bottom-up', inductive Aristotelian method. The instruments of investigation are limited to, and conditioned by, the five physical senses and their extensions such as telescopes and microscopes. The result is a precise description of the appearances, behaviour and physical mechanisms of the universe. The mind process is predominantly intellectual, applied in a linear, sequential mode.

It was Einstein who stated that there is no such thing as a 'crucial' or 'ultimate' test of a theory (or hypothesis) since all such tests assume the truth of other laws and theories in their formulation and execution'[6] The truth investigated by science is therefore relative and not absolute. At no stage can a scientist claim to have reached finality as any theory must be revised in the light of new facts.

Conversely, occult science works from universals to particulars within the limiting boundaries of evolutionary growth of every world system. The overall grand picture is first realised in its essential nature and the way this presents itself as particular effects is then expounded. The mind process is essentially holistic, applied in an all-inclusive mode. This is the 'top-down', deductive Platonic method. The instruments of investigation are *not* limited to the physical senses, which are *transcended* rather than *extended* as is the case with natural science. This gives profound insights into the origin, essential nature, manifestation, and expression of nature.

Natural science provides enormous factual information on individual physical phenomena but knows less about their intrinsic nature. In contrast, occult science reveals the grand plan but can sometimes provide inaccurate details about specific, vital physical mechanisms, which are best left to natural science.

Thus, the vital difference between the scientific and the occult methods of investigating nature and man is knowing from the outside and understanding from the inside, namely, knowing by detached observation and participation through union.

H. P. Blavatsky collected huge tracts of arcane literature from diverse epochs and cultures in order to underscore their common origin in the Mystery Teachings of antiquity and to demonstrate their highest common factor running as a thread binding them into a coherent whole. In her words: 'It is only by bringing before the reader an abundance of proofs, all tending to show that in every age, under every condition of civilization and knowledge, the educated classes of every nation made themselves the more or less faithful echoes of one identical system and its fundamental traditions [the Secret Doctrine, or Mystery Teachings]'.[7]

What are the litmus tests of truth? First and foremost, the self-consistency of doctrine concerning all departments of nature. Another is universality. Has it been taught by all those who have received the sunlight of initiation into nature's deeper secrets? So, for example, did Gautama, the Buddha in the East, instruct his disciples in the self-same doctrine that Jesus the Christ did in the West? Have countries such as India, ancient Persia, Egypt, Greece, China, ancient America, Iceland, Wales, and ancient Babylonia all received a wisdom-teaching which, stripped of outer vestments, is one in essentials? *Most assuredly, it is so*, for such patterns have been woven on one loom—the ageless loom of truth embedded in the Perennial Philosophy under names such as occultism, esotericism, or theosophy.

Throughout history, artists, musicians, poets, mystics, and lovers have known about the loss of self where the personal ego dissolves resulting in a state of subjective fusion with an idea, a subject of enquiry or an object of knowledge. Many scientists would like to pride themselves on their capacity to distance subject from object, claiming that such detachment results in 'pure' observation, free from personal bias. But scientists are no exception to the rule. The greatest scientific discoveries have come from those who were able to turn object into subject, so to say, that is, to participate actively in the inner state or consciousness of whatever they were investigating.

The technique known as *Samyama* is a yogic siddhi[8] (psychic power) 'in which subjects are able to learn about whatever they wish, without the use of intellectual intermediaries, by merging themselves with the chosen object.'[9] Physicist Stephen Phillips states that, 'the experience of a person in this state is not that of a passive spectator peering down a microscope. Instead, it is characterized by a vivid, subjective sense of actually being in the microcosm, of being suspended in space amid particles in great dynamical activity.'[10]

But there is no question whatsoever of science and occult science working in isolated, water-tight compartments. Each is reflected in the other. The inductive method of science includes the deductive method of occultism in the background.

Occult Science – an Overview

What facts do occult science legitimately and lawfully divulge? Occultism is in radical contrast to the mainstream scientific viewpoint that the universe is basically inanimate and that all life is a fortuitous concurrence of physical atoms with consciousness a by-product of purposeless physical processes and mind is a correlate of brain function.

Occult principles are the cumulative independent testimony of countless generations of sages of all times, all races, and all religions. As stated above, there is consistency in what they have declared in diverse tongues and forms from their own direct experience by incessant observations and rigorous independent corroboration. By 'direct experience', we mean strictly the *spiritual* visions and experiences attained after the most arduous and rigorous training and practice.

This occult doctrine received its clearest, most detailed, and forceful exposition from Blavatsky in her writings, notably *Isis Unveiled, The Secret Doctrine, The Key to Theosophy, The Voice of the Silence* and *The Collected Writings*. Collectively they have provided a unique service to humanity by drawing together the priceless but abstruse and far-flung flowers of a universal and eternal wisdom-tradition into one bouquet. Much to the point, her writings are replete with detailed source references and therefore *eminently verifiable*.

There are no conflicts or anomalies between the great occult traditions once the dogmatic interpretations and outer forms of expression are stripped out to reveal the bare inner meaning.

The depth and sophistication of occultism is prodigious. However, there is only a trio of key occult axioms given out as the *Three Fundamental Propositions* by Blavatsky in *The Secret Doctrine*[11] simplified below:

1. The Fundamental Unity of All Existence: an Absolute Principle of Unity behind diversity. The Causeless Cause behind all causes, unthinkable by human expression or comparisons, therefore ineffable and unpronounceable. 'It is that existence is ONE THING, not any collection of things linked together. Fundamentally there is ONE BEING.'[12]
2. Eternal Periodicity and Cyclicity: Universes appear and disappear from subjective realms to objective manifestation and from objectivity back into subjectivity. This fundamental law of the universe can be seen as the universal alternations of day and night, sleeping and waking or life and death.
3. The fundamental identity of all souls with the Universal Over-Soul: the latter an aspect of the Unknown Root; and the necessity for every Soul to realise and re-claim, in full consciousness, its intrinsic divinity. Every spark of life from the lowest to the highest seeks to be united with its parent source. So, there is involution (the progressive descent of spirit into matter) followed by evolution (the progressive ascent of spirit from matter). This means that there is a force in every creature that impels it to ascend towards ever higher forms of life. For example, the plant kingdom strives towards the higher consciousness of the

animal kingdom; the animal kingdom towards the self-consciousness of the human kingdom; and the human kingdom to superhuman realms of consciousness.

These Three Axioms can be expanded into Seven Principles:

1. An Absolute, incomprehensible (to the brain intellect) and Supreme Deity, or infinite essence as the root of all nature and all there is, visible and invisible, manifest and unmanifest, immanent and transcendent.
2. The hylozoic principle that says that there is no such thing in the universe as dead matter. Everything from below the mineral kingdom to beyond the human kingdom is conscious and interacts with, and responds to, its environment with intelligence and purpose. What varies, as we progress from sub-mineral to superhuman, is the degree, but not the fact of consciousness. The higher on the evolutionary ascent, the more awakened is the consciousness, because increasingly released from matter; the lower down the scale the more consciousness is dormant because progressively enmeshed in materiality. This is beautifully encapsulated in the proverb: I slept in the stone; I stirred in the plant; I dreamt in the animal; and I awoke in the man.
3. That man is a radiation of the Universal Soul and of an identical essence with it. The immortal nature of man is a spiritual entity and not the physical body its outermost vesture and vehicle on earth.
4. The Great Hermetic Axiom, often abbreviated, 'As Above, So Below'. This describes the relationship between the macrocosm—the universe as the great ordered whole—and the microcosm—man as its reflection. What this means is that the human mind is a microcosm of Divine

Mind. This involves the principles of analogy and correspondence, so that events having occurred in the external world are expressing operations and experiences of the human soul.

5. Theurgy, meaning 'divine work' or producing a work of the gods, is the application of esoteric principles and therefore an occult practice. By making oneself as pure as the divine, i.e., by reverting to one's pristine purity of nature, man can move the gods (i.e., command the forces and powers in nature) to impart to him divine mysteries.
6. The Law of Karma which adjusts effects to their impelling causes, both individually and collectively. This involves the doctrine of reincarnation but is not unique to mankind alone. It is a particular instance of periodicity and the eternal cyclicity of action and inaction, activity and rest, that may be observed in the natural world as the alternation of birth and growth, decay and death.
7. Evolution proceeds on three fronts: the spiritual, the intellectual and the physical. Consciousness ever seeks vehicles of increasing complexity, refinement and subtlety in order to express increasing amounts and degrees of its potentiality at infinite levels from the most spiritual to the gross material. It is clear that there is active intelligence behind the laws of nature. Concerning man, it affirms three cardinal principles about his evolution:
   i. New types, or species of mankind, have originated from more than one ancestral species.
   ii. There was a variety of modes of procreation before humanity fell into the ordinary (sexual) mode of generation.
   iii. The evolution of animals—of the mammalians at any rate—follows from that of man instead of preceding it.

The Occult Doctrine of Evolution

The second and third principles above are bound to raise many an eyebrow amongst Darwinian evolutionists. However, the evidence from occult sources is overwhelming and there is increasing evidence from palæontology. For example, a scholarly paper on the mating of Neanderthals and humans states that 'the skeleton of a 4-year-old child, recently unearthed at the 24,500-year-old site of Lagar Velho in Portugal, represents not merely a casual result of a Neanderthal/modern human mating, but rather is the product of several millennia of hybridization among members of the resident Neanderthal population and the invading *Homo sapiens*.'[13]

Recently, researchers in Germany discovered that the 50,000-year-old remains of a Neanderthal woman recovered from southern Siberia carried traces of *Homo sapiens* DNA. The analysis indicated that it was the result of one or more of her ancestors interbreeding with a human some 100,000 years ago. Whereas it is known that some people even today carry up to 4 per cent Neanderthal, the Siberian discovery shows, for the first time, that human DNA has been discovered in a Neanderthal.[14] By demonstrating the interbreeding and cross-breeding between different species with man these discoveries do not prove but allude to the logical possibility of the evolution of animals (mammalians) following rather than preceding that of man.

The term 'man' (derived from the Sanskrit word 'man' meaning 'to think') must therefore not be understood to be merely the physical creature with a body on two legs, with two arms and a head but rather as *a stage in the evolution of consciousness*. In other words, humanity should be regarded as a kingdom of nature just like the mineral, plant, and animal kingdoms. For this reason, man needs appropriate vehicles, other than his physical body, in order to manifest and actuate the vast range of his potentiality ranging from the physical to the spiritual via the intermediary of the emotional-mental characteristics of his soul.

Occult science does not reject the Darwinian theory so much as pointing out its severe limitations and incompleteness when it is proposed as the sole explanatory mechanism of evolution. The Welsh Nobel physicist Brian Josephson (*b.*1940) has said, 'There may be elements of intelligence in every atom of matter.'[15]

Neither matter nor mankind are thus the products of blind forces of nature or merely random evolutionary adaptations. On the contrary, both the visible and the subtle counterpart of the human body, down to the minutest speck of matter, are structured according to the cosmic blueprint or prototype of the 'image of God' referred to in mystical and religious traditions, such as in Christianity, Genesis 2:27. But the reader should *beware* that terms such as 'design', 'agents', and 'active intelligences' have absolutely nothing to do with the naïve and literal creed of 'intelligent design by a Creator-God' propounded by the Christian creationist-fundamentalists. They all point towards our transcendent origins—the divinity projected into and therefore mirrored in us.

'As Above, So Below' – the Hermetic Axiom

That which is below is like that which is above & that which is above is like that which is below.

The Hermetic Axiom is based upon the *Hermetica* or *Corpus Hermeticum,* the esoteric teachings of Hermes Trismegistus ('Thrice Great Hermes') who, 'whether as the Egyptian Thoth or the Greek Hermes, was the God of Wisdom with the Ancients and according to Plato, 'discovered numbers, geometry, astronomy and letters.'[16]

It states that the human mind is a mirror of Divine Mind. The core meaning of the Hermetic Axiom is that there always exists a correspondence between the laws and phenomena of the various planes of existence from the densest (the terrestrial) to the subtlest (the spiritual). There are planes of existence beyond our knowing since they exist in space-time orders outside the physical, but when we apply the principle of correspondence to them we are able to understand—by analogy—much that would otherwise remain unknown and unknowable to us. Thus, whatever happens on any level of reality (physical, emotional, mental or spiritual) also happens, correspondingly, on every other level. It is this principle that justifies the ancient sciences of astrology, alchemy, and magic by providing their theoretical basis.

The ever-presence of the deity/universe/man trinity and the recognition that there is absolutely no such thing as 'dead matter' would, if imported into the mainstream scientific ethos, resolve innumerable conundrums in cosmology, evolution, and the nature of consciousness.

The proper application of the Hermetic Axiom involves participating with living nature in all her forms. Its key themes are:

1. A complete absence of any fundamental dualism, whether between deity and his creations (including man) or between so-called good and evil.
2. The divine essence, and all of the universe and nature, is reflected, holographically, in our individual minds. The universe is a manifestation of deity.
3. We should therefore be interested in everything that is in the world. The particular and the concrete are the ways in which deity makes himself manifest to our experience.
4. Humanity spans the polarity from the spiritual to the material. Hence our minds can connect with intermediary spiritual intelligences to reawaken our dormant spiritual nature.
5. Based on the scenario of the fall into matter, the will, both divine and human, can be summoned to the regenerative work of redemptive reintegration with the divine and the associated refinement of Earth.

The Mystery Teachings About Man

Our physical body has a million billion cells—far more than the stars in the Milky Way. Of these 600 billion are dying and the same number are regenerating each day, which equals over 10 million cells per second. But no matter how diverse the cells, organs and physiological system, they act as one unity. We owe a tremendous debt of gratitude to medical science and neuroscience which have revealed so much of the wondrous workings of the physical body which is also a temple of the spirit.

Nonetheless, man has never ceased enquiring: What is *life*? What is *intelligence*? What is *force*? But we still have no satisfactory answers from mainstream science, physical medicine or orthodox religion on the profoundest questions about our existence.

In posing that most unfathomable riddle of all: *Who, or What Am I?* we are obviously not restricting man just to his physical body. However much microscopes and MRI scans may reveal the incredible mechanisms of the body, will these instruments of physical investigation open even a small chink into our inner or subtle bodies and higher states of consciousness?[17] For that inner vision we need other instruments of investigation: elevated faculties of consciousness.

The occult doctrines espoused in the Mystery Schools provide that necessary extra dimension to complement—not discard—the scientific picture. As the 'master rule' of occult investigation, we now trace the Hermetic Axiom's application in the Mystery Schools regarding man's true nature. These teachings of every age and epoch appear to be remarkably consistent in their various depictions about the composite nature of man. However varied and diverse the language, allegory or symbolism they invoked, and continue to invoke, the Hermetic Axiom.

What was the central purpose of all the Mystery Schools? They were established for the purpose of unfolding the nature of man according to certain universally tested and strict rules which, when faithfully followed, elevated the human consciousness to a point where it was capable of recognising its own constitution and the true purpose of its existence. The esoteric or occult doctrine is all about how man's multifaceted and composite constitution could be most quickly and completely regenerated to the point of spiritual illumination.

Greek and Early Christian Mysteries

The great Athenian philosopher Socrates maintained that the soul existed before the body and prior to its confinement therein was endowed with all knowledge. He held that when the soul entered the material form it became confounded, but by appropriate practices and virtuous living was able to struggle free of its impediments and so awaken to and recover its lost knowledge.

The Bacchic and Dionysiac Rites of the Greek Mystery Schools are based on the allegory of the youthful god Bacchus (or Dionysos) being torn to pieces by the giant Titans, who then scatted the dismembered body far and wide. Jupiter, the Father of Bacchus beholding this crime, hurled his thunderbolts at the Titans and slew them, burning their bodies to ashes with heavenly fire. From the ashes of the Titans, which also contained a portion of the Bacchic body that the Titans had partly devoured, the human race was created. Thus, the mundane life of every man was said to contain a part of the Bacchic and therefore godly life.

The rites taught that earthly man's composite nature comprised his lower nature consisting of fragments of the giant Titans and his higher, immortal, nature as the sacred life of the god Bacchus. So, the Bacchic state signifies the unity of the rational soul in a state of self-knowledge and the Titanic state, which by being scattered, loses the consciousness of its own essential one-ness or unity. The giants accomplished the fall of Bacchus by getting him to become fascinated with his own image in a mirror. Hence, ordinary man is capable of either a Bacchic—rational—existence, or a Titanic— irrational—existence.

In the Eleusinian Mysteries of the Greeks, the soul of man is essentially a spiritual entity having its true abode in the higher realms where, free from the bondage of material forms and concepts, it is said to be truly alive. The human, physical nature of man is likened to a tomb or bog and being a false, illusory and impermanent thing, constitutes the source of our suffering and sorrow. Indeed, Plato described the body as the sepulchre of the soul. He regarded man as constituted of two parts—one eternal, formed of the same essence as absoluteness, the other mortal and corruptible.

All this shows that the Greeks (as well as the early Christians) regarded man's nature as composite, requiring 'bodies' at various levels to express his entire nature from the material via the intermediary soul to the spiritual.

Egyptian Mysteries

The Greek Mystery Teachings were the offspring of the Egyptian Mystery Schools whose principal exponent was Thoth or the Greek Hermes Trismegistus. He wrote the *Corpus Hermeticum* which tells us:

Before the visible universe was formed its mold was cast. This mold was called Archetype, and this Archetype was in the Supreme Mind long before the process of creation began. Beholding the Archetypes, the Supreme Mind became enamoured with its own thoughts; so taking the Word as a mighty hammer, It gouged out caverns in primordial space [...] at the same time sowing in the newly fashioned bodies the seeds of living things [...]. Then the Father—the Supreme Mind—being Light and Life, fashioned a glorious Universal Man in Its own image, not an earthy man but a Heavenly Man dwelling in the Light of God. The Supreme Mind loved the Man It had fashioned and delivered to Him the control of the creations and workmanships [...]. The Man, looking into the depths, smiled, for He beheld a shadow upon the Earth and a likeness mirrored in the waters, which shadow and likeness were a reflection of Himself. The Man fell in love with His own shadow and desired to descend into it. Coincident with the desire, the Intelligent Thing [Man's high powers of mind and reason] united Itself with the unreasoning image or shape [illusory form]. Nature, beholding the descent, wrapped herself about the Man whom she loved, and the two were [ever] mingled. For this reason earthy man is composite. Within him is the Sky Man, immortal and beautiful; without is Nature, mortal and destructible.[18]

There are two points which are especially noteworthy. First, that man or rather 'Universal Man', was a thought form in 'Supreme Mind', obviously referring to the blueprint or archetype of man. Secondly, in addition to the teaching on the twofold composite nature of man, we find the clear message that sorrow and suffering are the consequence of immortal man (our higher part) falling narcissistically in love with his own shadow or image, mistaking his body and personality for his true self.

Another insight into man's physical, psychic, and spiritual natures comes from the symbolism attaching to the supreme god of Egypt, Osiris. As a solar deity, Osiris represents the material, life-giving aspect of solar activity. This vital principle—active, masculine—is the exact complement to the passive principle—inactive, feminine—represented by the deity under many names, notably Isis, signifying the principle of natural fecundity.

Osiris symbolises the dual (composite) nature of mankind: the cosmo-spiritual and the terrestrial, or the divine and the physical human. Osiris is rather curiously also identified as the god of the afterlife, the underworld, and the dead.

Persian Mysteries

The Persian Mysteries incorporated the rites of the sun-god Mithras who has both a male and female aspect. As *Mithras,* he represents the masculine principle, being Lord of the Sun, radiant and powerful; as *Mithra,* this deity represents the feminine principle, nature as receptive and terrestrial. Mithra is fruitful and productive only when conjoined with the solar fire of Mithras. Mithras also stands for the god of intelligence, mediating in the struggle for supremacy between Ahura-Mazda, or Ormuzd, the spirit of good, and Ahriman, initially a pure spirit who then rebelled against Ormuzd, being envious of his power.

So when Ormuzd created the Earth, Ahriman entered into its gross elements and whenever Ormuzd performed a good deed, Ahriman planted a seed of evil within it. Finally, when Ormuzd fashioned the human race, Ahriman became incarnate into the lower nature of man such that in every human personality there is a constant war between the spirit of good and the spirit of evil, each struggling for control.

Let it be said that the wisdom-religion-philosophy of Persia, as indeed of Greece Egypt, India, and other great centres of esoteric learning, are united in affirming that there is no *independent* principle of an 'Evil One' by itself—by whatever name we might care to call that shadow, e.g. Ahriman, Devil, Satan, Diabolos or Mara, or the Egyptian 'Lord of Chaos', the serpent Apep/Apophis, who embodied darkness and disorder, and was thus the opponent of light and truth. Certainly, there have been evil individuals and groups—we hardly need any convincing about that. In such cases it may be said that, apparently, the Satanic principle has completely overpowered and controlled the benevolent principle.

Nonetheless, there is no metaphysical principle of evil *per se.* Good and evil, so-called, therefore come into existence only when man in his onward evolutionary march develops the power of *knowledge* and *choice.*[19]

The Catholic Church has ingeniously anthropomorphised Adam and Eve in the Garden of Eden. The Devil (Satan) has also suffered similar ludicrous treatment in the hands of those who mouth sermons without understanding their esoteric import. It is amazing that even today highly intelligent people attribute a literal meaning to the allegory of Adam and Eve.

Indian Mysteries

Ancient India had, arguably, the most diverse, complex and richest of all the Mystery Schools and occult teachings that have provided the wellspring of the later Mystery-centres of Egypt and the West. Given the highly individualistic character and independence of the Indian mind and the resultant tendency to doubt, challenge and question everything in religion and philosophy, it is not altogether surprising that two great streams of divergent thought emerged from a single source: the Vedanta philosophy and the Samkhya philosophy, representing the manifest tension between revelation and reason.

Samkhya philosophy was propounded around the seventh or eighth century BC. It accepted the Vedas as canons of knowledge (but discarded, or ignored, many of its theories and ideas) and was also influenced by Buddhism and Jainism. Samkhya philosophy propounded universal evolution out of root matter, or primordial nature (*prakriti*), as the source of all manifestation, and postulated an infinite number of immortal souls *purusha*. Its reputed founder was the sage Kapila, who propounded an atomic theory as part of his philosophical speculations on universal evolution, interestingly, along broadly similar lines, and during the same era, as the ancient Greek philosophers, principally Democritus (*circa* 460–*circa* 370 BC), Plato, and Aristotle.

In the much older Vedic tradition, the Mysteries were approached and taught from two major standpoints, as indeed was the case in all Mystery Schools. The Lesser Mysteries were given out openly, albeit entirely in allegorical and symbolical format. These were the Puranas, meaning 'belonging to olden times', now eighteen in number and containing the entire body of ancient Indian mythology. The underlying teaching of the Puranas concerned the general sequence and process of creation and the forces in nature, intelligent and semi-intelligent, disseminated in epic stories about the adventures of various male and female deities, such as the story of Krishna, the greatest of the avatars. These teachings can broadly be classified into five themes:

1. Cosmogony, i.e., the beginnings and 'creation' of the universe.
2. Periodical renewals and destructions of universes.[20]
3. The genealogies of the gods, other divine beings, heroes and patriarchs.
4. The reigns of the various Manus (divine beings representative of the perfect Man).
5. A résumé of the history of the solar and lunar Races.

The Puranas were supposed to have been composed by the poet-sage Vyasa, the author of the legendary *Mahabharata*. In the *Bhagavad Gīta* (the crown jewel of the *Mahabharata*), the Lord Krishna symbolises the Christos principle that unfailingly brings man to the state of true individual perfection by teaching the human intelligence (represented by Arjuna) how to acknowledge, aspire to and activate its highest and innermost principle.

In contrast, the sacred knowledge in the Vedas, meaning 'divine knowledge' (compiled by the sage Veda-Vyasa), was accessible only to the initiated Brahmins for their individual spiritual training and instruction. The four Vedas are concerned with the deepest aspects of esotericism and occult philosophy. They represent the best Sanskrit compendium of ancient sacred laws and customs of divine origin and remotest antiquity. The Upanishads, usually reckoned today as one hundred and fifty in number, are part of the Vedic cycle constituting its esoteric doctrine—the Greater Mysteries.

The topics are highly transcendental, recondite and abstruse. The Vedanta (perhaps the noblest of the six Indian schools of philosophy) is a mystic system of philosophy that was developed from the efforts of generations of sages to interpret the secret meaning of the Upanishads. So, we can trace a noble current of mysticism and esoteric philosophy from the Upanishads, Vedanta and other lofty systems of Indian philosophy—all arising from, and rooted in, the parent Vedas.

American Mysteries

The Mysteries of the Americas underscore their commonality with other Mystery-centres of the world. Their underlying philosophies apply, in a general sense, to both American continents. It is important to distinguish between the completely different goals of the colonisation of North America and the conquest of Central and South America. The European immigrants who went to North America were first the Huguenots (French Protestants) in the middle sixteenth century, followed in the seventeenth century by the British. Their intention was to build a new, free country based on the principles of Rosicrucianism and Freemasonry. Although they were not free of blame for having taken the native lands by force and disbanding, or annihilating, many tribes, their primary aim was to colonise the continent and seek new opportunities for prosperity and religious freedom.

The Spaniards discovered the wealth of the Incas in the 1520s and went to Central and South America with the express purpose of plundering the Incan Empire of its wealth, ransacking its natural resources and subjugating the populations by crushing their cultural and religious traditions.[21] The conquest took place under the guise of religious conversion as the Spanish Inquisition regarded the religious beliefs and practices of the

Inca, Maya, and Aztec peoples as profane and blasphemous to Catholicism.

North America

This section is based mainly on the exposition by Manly Hall[22] and other sources. The Mysteries of North America were the outcome of the nature of its peoples, innately drawn to symbolism and mystical philosophy. Like the aboriginal peoples of Australia, Africa, and elsewhere, the Indigenous Americans' animist instincts about elemental creatures resulted from their intimate contact with nature. Thus, the whole sky, rivers, and forests were inhabited with myriads of superphysical and invisible beings. Their mystical philosophy had much in common with other world cultures. The North American Indians (Native Americans) considered the Earth (the Great Mother) to be an intermediate plane, bounded above by a heavenly sphere (the dwelling place of the Great Spirit) and below by a dark and terrifying subterranean world (the abode of shadows and of sub-mundane powers).

They divided the interval between the surface of Earth and heaven into various strata, one consisting of clouds, another of the paths of the heavenly bodies and so on. The underworld was similarly divided. Those creatures capable of functioning in two or more elements were considered as messengers between the spirits of these various planes. The abode of the dead was presumed to be in the heavens above, the Earth below, the distant corners of the world or across wide seas. Sometimes a river flowed between the world of the dead and that of the living.

The legendary narratives of the strange adventures of intrepid heroes who, while in the physical body, penetrated the realms of the dead prove beyond question the presence of Mystery cults among the indigenous peoples of North America. Wherever the Mysteries were established they were recognised as the philosophic equivalents of death, for those passing through the rituals experienced all after-death conditions while still in the physical body. At the consummation of the ritual the initiate actually gained the ability to pass in and out of his physical body at will.

The North American Indians recognised the difference between the ghost and the actual soul of a dead person, a knowledge restricted to initiates of the Mysteries. They also understood the principles of an archetypal sphere wherein exists the patterns of all forms manifesting on the Earth plane. The theory of 'Group', or 'Elder Souls' having supervision over the animal species is also shared by them.

As Hall avers, 'who can doubt the presence of the secret doctrine in the Americas when he gazes upon the great serpent mound in Adams County, Ohio, where the huge reptile is represented as disgorging the Egg of Existence?'[23]

The Mysteries of the North American Indian had much in common with the mystical philosophy of other world cultures. Their view of the Earth as an intermediate plane, bounded above by a heavenly sphere and below by a subterranean world, was similar to the cosmovision of the early Scandinavians. Like the Chaldeans, the interval between the surface of the Earth and heaven was divided into various strata and like the Greeks, the underworld was similarly divided. The river that flowed between the world of the dead and that of the living, paralleled the Egyptian, Greek, and Christian theologies. Their understanding of archetypes, as the universal patterns of all pre-existing forms that manifest on Earth, was in common with the ancient Platonists.

Central America

The *Popol Vuh* is the story of creation according to the advanced civilization known as the Quiché (K'iche') Maya of the region recognized today as Guatemala. No other sacred book, according to Hall, sets forth so completely the initiatory rituals of a great school of mystical philosophy. This volume alone is sufficient to establish incontestably the philosophical excellence of the peoples indigenous to Central America.[24]

The American author and Theosophist James Morgan Pryse (1859–1942) approached the *Popol Vuh* from the standpoint of the mystic, calling this work *The Book of the Azure Veil*. 'The Red "Children of the Sun" do not worship the One God. For them that One God is absolutely impersonal, and all the Forces emanated from that One God are personal.'

Note—this is the true mystic and esoteric philosophy of all ages. It is the exact reverse of the popular Western conception of a personal (even anthropomorphised) God and impersonal forces in nature working according to mindless mechanical laws—the cause of no end of confusion and strife in orthodox religion and establishment science.

The Xibalbian Mysteries are found in the second book of the *Popol Vuh,* which is largely devoted to the initiatory rituals of the Quiché nation. These ceremonials are of seminal importance to students of Masonic symbolism and mystical philosophy since they establish, beyond doubt, the existence of ancient and divinely instituted Mystery schools on the American Continent.

As with the initiatory trials in any Mystery School, the Xibalbian processes were intended to test, to the very limit, the character and mettle of a candidate before he could be entrusted with the innermost secrets of nature. An example of just one of the ordeals: Hunahpu and Xbalanque (the heroes of the second book of the *Popul Vuh*), are depicted undergoing the torment of the Bat House. Camazotz, the Lord of the Bats, emerging suddenly from the gloom, strikes with his great sword at the intrepid invaders of his domain. The test here is the realisation that the animal soul of man is likened to a bat because, like this creature, it is blind and can only be restored to sight by the light of the spiritual, or philosophic sun—the spiritual soul.

Then, as time passed, we learn from the available native records the abundant evidence that the priestcraft of later civilizations of Central and South America degenerated into black magic. A similar trend that can clearly be perceived elsewhere, such as in the latter days of the Egyptian dynasties and in India. The abominable sanguinary rites practised by many of the Central Americans possibly represent remnants of the Atlantean perversion of the ancient sun Mysteries. The secret tradition submits that during the later Atlantean epoch black magic and sorcery overpowered the esoteric schools, resulting in the bloody sacrificial rites and gruesome idolatry which ultimately overthrew the Atlantean empire and even penetrated the Aryan world, fragments of which may be discerned even to this day—recall the Nazi atrocities and the recent abominations of the Taliban and the Islamic State in Iraq and Syria.

South America

It is well said that the history of a nation is written by its conquerors. This certainly applies to Latin American history written under the censorship of the Spanish Inquisition, which spanned nearly four centuries from 1478 to 1834. Much of the recorded information about the great Inca, Maya, and Aztec civilizations and empires was therefore distorted. But it can be recovered through its myths, legends, chronicles, and fragmentary documents left behind by the Spanish invaders in the sixteenth century, and the tales of the mestizos (persons of mixed racial or ethnic ancestry), as well as archæology and anthropology. We find scattered amongst these sources, hints and traces of information about the real dimension of the pre-Inca and Inca worldview.[25] There appears to be a persistent theory that the Incas embodied the final phases of a highly evolved culture which started around the eleventh century.[26]

In *The Cosmovision of the Incas*[27] the contemporary Latin-American sociologist, Dr Albert Soria informs us that a highly organised state religion existed in the Inca Empire. The Inca priests resided at important shrines and temples. The official priesthood maintained temples and convents in the main cities of the Quechua people (one of the indigenous races of South America). The Sun god, known as 'Inti', was the most important deity and is usually portrayed as a gold disc with a human face from which rays and flames extended. The second most important deity was the Moon, 'Mama Killa', who was Inti's sister and consort, represented as a silver disc with human features. Their offspring were the stars and thunder. Then came the Earth Mother, 'Pachamama'. She was regarded as a living entity and mother of all earthly life, plus the personification of fertility and growth.

The common characteristic in the mythologies among the Inca, Maya, and Aztec civilizations is their accounts of supernatural beings coming out of water—a lake or sea. The rituals of these people indicate that they worshipped the Sun. The Incas considered themselves Children of the Sun.

The American metaphysical author George Hunt Williamson (1926–1986, pen name 'Brother Philip') writes about the legend of a highly evolved civilization that was destroyed by a series of cataclysms around 11,500 years ago. (There is considerable scientific evidence for worldwide cataclysms during this time.[28]) Survivors of these disasters migrated to different parts of the world, some moving to the Andean Cordillera through the Eastern Islands of the Pacific. The legend further relates that an initiate by the name of Amaru Muru—from whose name 'America' is derived—established a highly spiritual society on the shores of Lake Titicaca at Puno in Peru. This was the Brotherhood of the Seven Rays which had become the leading force in the spiritual lives of the Incas.[29]

According to the French philosopher Serge Raynaud de la Ferrière (1916–1962), the Andean Cordillera has a special regenerative energy for the Americas such as Tibet used to have, and possibly still does, for Asia. He refers to the safety of an old Inca temple, a Mystery-centre in the Andean Plateaux as a 'sacred place where the profane have never come, in spite of the different organized expeditions.'[30] In the same way that Tibet and India were once the focus for pilgrims in search of enlightenment, the Andes is now entrusted with this function in the coming age.[31] The American spiritualist Earlyne Chaney (1916–1997), spiritual leader of the esoteric group *Astara* writes in *Lost Empire of the Gods*: 'The old Peruvian said that in these lost inner-world cities, hidden behind the cordilleras of the Andes, dwell these emissaries from an ancient civilized race.'[32]

In the classic volume *Sacred Mysteries among the Mayas and the Quichés* the French-American photographer, archaeologist, and antiquarian Augustus Le Plongeon (1825–1908) compared the mythology and spiritual belief systems of the Mayans and those of other cultures, such as the ancient Egyptians, Indians, Chaldeans and on to the Christians. He illustrates how entirely human many sacred mysteries are and just how closely bonded people and communities are—no matter how far apart.

In *The Cosmovision of the Incas* Albert Soria argues that esoteric writings suggest there was a spiritual link between the Incas and the Western Mystery Tradition—indeed, one that may still exist.[27] Among these Traditions are the Alpha and Omega, the Order of Melchizedek and the Order of the Essenes, as well as, in South America, the Brotherhood of the Seven Rays, and the Solar Brotherhood located in Urubamba Valley, the Sacred Valley of the Incas, near Cusco in Peru. Regarding the connection between the Incas and the Orders comprising the Great White Brotherhood, dispersed all over the world, it would appear that there was some sort of contact between the mystical organisations of Europe and the secret societies of the Andes.

It is evident that the universal, esoteric, and mystical tradition, from its primeval origins in the mists of antiquity, has spread out to fecundate Mystery-centres around the world, principally in Asia, Egypt, Greece, Europe, and the Americas.

Very little is known or revealed about the mystical orders that reportedly dwell in the fastness of the Himalayas and the Andes or in other remote regions of the world. This is entirely as it should be, for authentic spiritual societies have no need for marketing, self-promotion or public recognition. Their teachings are generally orally transmitted to qualified and genuine seekers and, apparently, the first contact is usually made on levels that are non-physical (astral). That is the way to contact such as the True and Invisible Rosicrucian Order in Europe, the mystical Order of the Andes or the Mahatmas of India.

Contemporary Mystery Teachings

Modern centres of Mystery Teachings are needed now more than ever due to the present turbulent world situation. There are, apparently, three principal inter-related distributions of this esoteric life-flow from the heart-centre of Shamballa through the Lesser and Greater Mysteries.[33]

The first, as the American scholar and esoteric philosopher Gottfried de Purucker (1874–1942) informs us, comprise the Lesser Mysteries that are now 'largely replaced by the different activities of the Theosophical Movement which itself is exoteric as a Movement'.[34] Far more carefully concealed because of excessive materialism in science and philosophy blinding humanity to spiritual insights, are the Greater Mysteries comprising the Esoteric Section of the Theosophical Society and the still more recondite Inner Group Instructions of the Esoteric Section. But note, that there are no grounds for assuming that the Greater and Lesser Mysteries are the sole prerogative of the Theosophical (or Anthroposophical) movements.

The second 'distribution centre' is through the spiritual centres of the nations. Since all countries are in magnetic and sympathetic vibration with Shamballa, these act as organic focal points for the circulation of spiritual influences. Thus, every country has its own secret spiritual protectors who as a body form a true esoteric centre. It cannot be too strongly emphasised that these national 'occult guardians' do not meddle in political affairs. As Purucker states, their work is 'purely spiritual, moral, intellectual, and wholly benevolent, and indeed universal, and is a silent guide to the intuitive minds of the different races.'[35]

The third channel of esoteric work is one of the most fascinating but least recognised. Its aim is to preserve the knowledge from age to age. Purucker said: 'There are actually groups whose sole business is forming occult centres of Initiation, preparing of students for esoteric work in the world, and for the safeguarding of priceless treasures, the heirlooms of the human race, treasures both intellectual and material.'[36]

According to Purucker, the chief of these veiled centres have branches in Syria, Mexico, Egypt, the United States, and Europe, each one 'subordinate to the mother-group of the Occult Hierarchy in Shamballa.'[37] At first sight, it might seem incongruous that a country like war-torn Syria would have an esoteric centre of spirituality. But it is precisely the darkness of ignorance and stupidity that calls forth a compensating centre of truth and illumination as an indispensable agency. And in any case, some of the world's greatest masters were born in ancient Syria—Pythagoras and Jesus, to name but two.

It appears, then, that all that is of essential spiritual value is preserved in the secret archives of the planet, for the generations of seers are not wasteful, nor are the grand systems of philosophy and religion lost in the darkness of receding ages. As Blavatsky avers: 'There are, scattered throughout the world, a handful of thoughtful and solitary students, who pass their lives in obscurity, far from the rumors of the world, studying the great problems of the physical and spiritual universes. They have their secret records in which are preserved the fruits of the scholastic labors of the long line of recluses whose successors they are.'[38]

Someday, worthy explorers will recover the lost keys, the Mystery Teachings and the secrets of the human race such as are allegedly concealed in the secret chambers of the Great Pyramid. As spoken (telepathically) by one of the Guardian spirits of the Great Pyramid, as recorded by Paul Brunton: 'In this ancient fane lies the lost record of the early races of man and of the Covenant which they made with their Creator through the first of His great prophets. Know, too, that chosen men were brought here of old to be shown this Covenant that they might return to their fellows and keep the great secret alive.'[39]

When will this happen? Not by man's whim or fancy, but at the appointed hour when an awakened intuition reveals the light of perennial truths.

The Mysterious Shamballa

As expounded by the international exponent of Vedic philosophy, the Hindu-American David Frawley (*b.*1950): 'According to the Tibetan Buddhists, there was an earlier Buddhist kingdom in Central Asia from which their teaching derives, called Shamballa, which appears to have been in the Tarim basin, northwest of Tibet [...]. This earlier Buddhist culture was, by Tibetan accounts, also a culture in which [an earlier, original version of] Sanskrit, the "language of the gods", was spoken.'[40]

Here, according to the most ancient traditions, lived those great intelligences (*Kumaras*) who are the real guides of our planetary life as a whole, keeping watch over the evolution of the various kingdoms of our planetary nature, seen and unseen—not just mankind. These intelligences were known in antiquity as 'Lhas', from which, apparently, the Islamic god-name 'Allah' is derived. According to occult sources, these same intelligences are those who have, several aeons ago, passed beyond the range of human existence into the superhuman realm known as 'Planetary Spirits'.

The same high sources inform us that there was a time, several million years ago, when at least some of the Shamballic inhabitants took on physical bodies as their outer vestures and moved openly among the prototypal humanity of the time (the Third, or Lemurian Root Race) in order to assist with the nurturing of the full human type.[41, 42] This was at a time when, apparently, an earlier equivalent of Shamballa existed in the area of what is now central South America.

Once the full individualisation of self-conscious man had been achieved, the location of Shamballa moved to the other side of the world to await and assist the development of the next great racial types—the Fourth or Atlantean Root Race and the Fifth or Aryan Root Race. The latter, also known as the Indo-Caucasians, is the racial type of the vast majority of present-day humanity. At some intermediary stage, the objectively visible land of Shamballa and its inhabitants transferred to a higher state of being, which, to ordinary mortal sight, is invisible. And it is for this reason that modern travellers or archaeologists have found no physical or objective trace of it. This sphere of operation in unseen realms of the Shamballic guardianship and evolutionary guidance is confirmed in the cryptic writings of occult fraternities such as the Rosicrucians.

In the same way as a family or a corporation, a kingdom or an empire has its organisational structure, Shamballa is organised with three superior (unseen) Kumaras governing the higher kingdoms of our planetary nature. With the assistance of devas (the 'shining ones' from a kingdom above the human) four other Kumaras govern the other kingdoms of nature.

The seventh Kumara governs the mineral kingdom and element of earth.

The sixth, the plant kingdom and the element of water.

The fifth, the animal kingdom and the element of air.

The fourth (Sanat Kumara), the human kingdom and man's higher principles via the adept hierarchy and the lower terrestrial nature via the devas and the element of fire.[43]

Related Nineteenth-Century Spiritual Teachings

New Thought and its precursor, Transcendentalism, are not in the same sublime category as the Mystery Teachings. But these philosophic-religious nineteenth century movements in the United States prepared the esoteric groundwork for the emergence of the complex occult doctrines promulgated through the Western occult tradition. These were organisations like the Hermetic Order of the Golden Dawn and the Theosophical Society also founded in the United States in the late nineteenth century. It is a matter of pure speculation whether they were in any way subliminally influenced by the Mystery Schools.

Transcendentalism was a major American intellectual movement and inspired succeeding generations of American intellectuals and literary notables.[44] (However, it was certainly not the first of its kind. For example, as the French occult author and ceremonial magician Éliphas Lévi (1810–1875) has noted, the Italian visionary mystic Comte di Cagliostro (1743–1795) was believed to have been an emissary of the Knights Templar, a late Middle Age society in Europe with strong transcendentalist tenets.[45]) The movement in America directly influenced the growing movement of 'Mental Sciences' of the mid-nineteenth century, which later became known as the 'New Thought movement' of which Ralph Waldo Emerson (1803–1882) is widely regarded as its philosophical and intellectual father.[46]

The aim of Transcendentalists was to base their religion and philosophy on principles drawn from the inner spiritual and mental essence of the person and not based on, or contradicted by, physical experience. The principal influences were the biblical criticisms of the German philosophers and theologians

Gottfried von Herder (1744–1803) and Friedrich Daniel Ernst Schleiermacher (1768–1834), and the scepticism of the Scottish philosopher David Hume (1711–1776) along with German and especially English Romanticism, the mystical spiritualism of Emanuel Swedenborg and Indian sacred literature.[47]

The 'New Thought' movement was, and still is, a spiritual movement with an emphasis on mind-healing that originated in the United States in the 1830s.[48] It originated from the overall milieu of transcendentalism and therefore has similar religious and philosophical ideals and precepts. It was based on religious and metaphysical ideas and prepared the ground for the profound doctrines promulgated through the Theosophical Society.

Its influence survives to this day in the form of a loosely allied group of spiritual and religious denominations, writers and philosophers, not only in the United States, but also in the United Kingdom, Europe, Asia, Africa, and Australia. Common features include metaphysics, positive thinking, the law of attraction, healing, the life force, creative visualisation, and personal power. This demonstrates that teachings concerning man's relationship with deity, cosmos, and nature are not confined solely to the Mystery Schools of antiquity or their modern versions.

New Thought's origins may also be traced to the romanticism and idealism of the nineteenth century combined with a growing dissatisfaction with scientific materialism and empiricism. It was also a reaction to the religious scepticism of the seventeenth and eighteenth centuries.

The American spiritual teacher Phineas Quimby (1802–1866) is usually cited as the earliest promoter of the movement. He developed his concepts of mental and spiritual healing and health based on the view that illness is a matter of the mind. His influence may be seen in the writings of Mary Baker Eddy (1821–1910), the founder of Christian Science.

New Thought is essentially positive and optimistic about life and its outcomes. The movement does not maintain any dogma, or just one belief system, apart from individuals like Baker Eddy who opposed medical science. It promotes the idea that the divine is omniscient and omnipresent, transcendent and immanent in everyone and everything. It holds that infinite intelligence is ubiquitous, spirit is the totality of real things, true human self-hood is divine, divine thought is a force for good, sickness originates in the mind, and right thinking has a healing effect. It is open to all religions, since spiritual healing and strength of mind and body are available to all who live their lives and marshal their thoughts according to nature's universal laws.

The Debasement of New Thought

More recently leaders of the movement have increasingly stressed material prosperity as one result of New Thought. Many individuals and groups have, regrettably, seen it that way. So, the core teachings have degenerated into the popular modern egoistical and highly commercial, self-glorification culture of trying to get just what one desires— invariably power, sensual gratification, riches, and celebrity status—simply by wishfully 'thinking the right way' or chanting sacred words but with no attendant self-effort.

The 'gospel of the prosperity preachers' now in America is deplorable. One in five Americans are estimated to follow a prosperity gospel church led by maestros of high-tech religious marketing—like Joel Osteen, whose personal fortune is estimated at $60 million and whose mansion with parking for twenty cars is valued at $10.7 million.[49] Jesse Duplantis, the American televangelist, is asking his followers to help him acquire his fourth private jet—a Falcon 7X tri-jet for $54m—because God had told him to buy one.[50] Latest in this line of televangelists is Paula White, spiritual adviser to former President Donald Trump, an appointment she calls an 'assignment from God'. Claiming that the Lord wanted her to go on national television, she practices the school of thought known as 'prosperity theology'.[51]

Ralph Waldo Emerson and the Over-Soul

Emerson was the cardinal exponent of Transcendentalism and New Thought and a supreme example, too, of what is called higher thought. Emerson's entire output—but especially his three celebrated and widely read works: *Compensation, Self-Reliance* and *The Over-Soul*—strongly presage Blavatsky's 'Three Fundamental Propositions' mentioned earlier. (Note: *The Secret Doctrine* was published six years after Emerson died.)

Emerson realised that when we try to describe God in words, both language and thought desert us since the ineffable essence refuses to be recorded. 'Of this pure nature every man is at some time sensible. Language cannot paint it with colours. It is too subtle. It is undefinable, unmeasurable, but we know that it pervades and contains us.'[52]

'What is the aboriginal Self, on which a universal reliance may be grounded?' asks Emerson.[53] This single source from which all human beings emerged is none other than the Over-Soul, as elucidated in arguably his finest work by that name. The four main themes are: (*a*) the existence and nature of the human soul; (*b*) the relationship between the soul and the personal ego; (*c*) the relationship of one human soul to another; and (*d*) the relationship of the human soul to God, i.e., how the Over-Soul manifests in individuals.

He described the Over-Soul as, 'that great nature in which we rest, as the earth lies in the soft arms of the atmosphere; that Unity, that Over-soul, within which every man's particular being is contained and made one with all other.' 'All is One' he repeats again and again, 'the act of seeing, the thing seen, the seer and the spectacle, the subject and the object.'[54]

However, he admits that ultimately *it cannot be known through language and definitions, but only through moral actions.*

Emerson concurs with the First Fundamental Proposition stated in the Proem of Blavatsky's *The Secret Doctrine,* that of basic Unity, tracing the movement from within to without, the One becoming many. In this way he affirmed the absolute necessity of both unity and diversity. 'Every chemical substance, every plant, every animal in its growth', he writes, 'teaches the Unity of Cause, the variety of appearance.'[55]

The various substances out of which nature's forms are compounded might seem to be divided yet they are all united. 'The Same, the Same' he declares; 'friend and foe are of one stuff; the ploughman, the plough and the furrow are of one stuff; and the stuff is such, and so much, that the variations of form are unimportant.'[56]

The Second Fundamental Proposition speaks of eternal periodicity and cyclicity. In a speech to the Phi Beta Kappa Society of Harvard College Emerson said: 'There is never a beginning, there is never an end, to the inexplicable continuity of this web of God, but always *circular power returning into itself* [writer's emphasis]. Therein it resembles his own spirit, whose beginning, whose ending, he never can find.'

Emerson's insight into universal periodicity paved the way for his understanding that the whole of manifested nature is pervaded by an inexorable duality so that each thing perceived is only one half and demands another half to make it a whole. He continues in the same address: 'Polarity, or action and reaction, we meet in every part of nature; in darkness and light; in heat and cold; in the ebb and flow of waters; in male and female; in the inspiration and expiration of plants and animals; in the systole and diastole of the heart; in the centrifugal and centripetal gravity; in the undulations of fluids and of sound; in electricity, galvanism and chemical affinity.'[57]

Emerson was fully aware of periodicity, as manifested through the Law of Karma, when he states, 'What we call Retribution is the universal necessity by which the whole appears whenever a part appears. The causal retribution is in the thing, and is seen by the soul. The retribution in the circumstance is seen by the understanding; it is inseparable from the thing, but is often spread over a long time, and so does not become distinct until after many years. Cause and effect, means and end, seed and fruit, cannot be severed; for the effect already blooms in the cause, the fruit in the seed.'[58]

The wise man, he counsels, will extend this lesson to every department of his life, and realise that *it is the part of wisdom to pay his debts on whatever plane*. Persons and events may seem to stand for a time between a man and justice, but this is only a postponement, for sooner or later the debt must be settled.

The Third Fundamental Proposition proclaims the fundamental identity of all souls with the Universal Over-Soul, like the sparks from a central fire. Emerson continually asserted that by ignoring spirit, science could never reach ultimate truth and that religion, by limiting itself to spirit and ignoring matter, was in the same position. He saw that something was needed which took both into account and offered a basis of reconciliation. This basis was his overall philosophy which he bequeathed to the world. He turned again to the East for a corroboration of scientific theories and prophesied that, 'The avatars of Brahma will presently be text-books of natural history.'[59]

What does Emerson explicitly have to say about man and his soul? In his essay *The Over-Soul*, he again clearly states his conviction that, 'the soul in man is not an organ, but animates and exercises all organs; is not a function, like the power of memory, of calculation, of comparison, but uses these as hands and feet; is not a faculty, but a *light*; is not the intellect or the will, but the *master* of the intellect and the will; is the vast background in which they lie, an immensity not possessed and that cannot be possessed [writer's emphases].'[60]

Clearly, then, he regards the soul as that permanent principle in man which does not change, but which is able to perceive the changes going on around him.

The composite nature of man—one part divine and immortal, the other part mortal and corruptible—did not escape Emerson's acuity. For he warns us, '[to] hold fast to the man and awe the beast; stop the ebb of thy soul—ebbing downward into the forms into whose habits thou hast now for many years slid.'[61]

Reincarnation figures strongly in Emerson's thoughts on the basis that he considered the soul as an evolving, *becoming* entity. Its advances, he says, are not made by gradation, such as can be expressed by motion in a linear fashion, but rather by an ascension of state (namely, higher planes of consciousness). There are innumerable steps on the stairway of evolution, he explains, which we have already climbed. But there are steps above us, many a one, which go upward and out of sight. How can these steps be climbed other than through the process of reincarnation?

In terms of the soul's unfoldment, it is apparent that Emerson considered this idea as the only logical one. He says: 'The soul having often been born, having beheld the things that are here, those which are in heaven and those which are beneath, there is nothing of which she has not gained the knowledge. No wonder that she is able to recollect [re- collect] what formerly she knew.'[62]

The resounding call for self-reliance supported by the divine power rings through Emerson's entire philosophy and he promulgated it throughout his life.

His aims and purpose were essentially similar to the Three Objects of the Theosophical Society in that he constantly reiterated the brotherhood of humanity, its First Object. He always encouraged the study of comparative religions, sciences, and philosophies as per the Second Object. And he continually pointed to the spiritual powers latent in man and urged their development, as per the Third Object.

Mystery Teachings and Nineteenth-Century Thought

The reader should never be under the impression that such emphasis on the primacy of thought was the province of gullible nineteenth-century make-believe without scientific justification. It has every scientific support. It is worth pointing out that everything that the reader can see looking around the place in which he is reading was once a thought—as indeed was this book! So, the unequivocal conclusion is the primacy of thought. This has been the doctrine of arcane wisdom transmitted through the great sages and occultists since time immemorial.

The Mystery Teachings never teach that spirituality ever makes men either virtuous, rational or compassionate. On the contrary, that virtue, rationality, and compassionate service rendered are the qualities that, in due course, make men spiritual. Accordingly, the Mysteries teach that spiritual illumination is attained only by raising the lower nature towards the highest grade of perfected functioning and purity. This knowledge of how man's composite nature can be most speedily and completely unfolded and regenerated to the point of spiritual illumination constitutes the secret, or esoteric doctrine, of all ages and is the main objective of the Mysteries.

Through the gradual purification of his vehicles and the ever-increasing sensitiveness resulting from that refinement, man, by gradual degrees, overcomes the limitations of matter. The secret doctrine of all ages further teaches that after aeons of incarnations on Earth, the evolutionary cycle for our present humanity will have completed its physical evolution and man will have attained his own essential selfhood. Thereafter the empty shell of materiality left behind will be used by other life waves as stepping-stones to their own evolutionary cycle.

However, those desiring to do so may enormously accelerate the natural evolutionary process by determined self-effort. These are the initiates and earnest students of occultism who have decided to take their spiritual development into their own hands instead of leaving it to the natural course of events. By initiation into the Mysteries they learn that man becomes aware of, and consciously reunited with, the divine source of himself without tasting of physical dissolution. This is at once the primary purpose and the consummate achievement of the Mysteries.

One of the best modern, authentic accounts of such an initiation is the experience of Paul Brunton during an entire night in the darkness of the King's Chamber of the Great Pyramid. The process commenced with a severe and terrifying psychic ordeal to test his mettle and resolve before the climax, when Brunton was given actual proof about death, the soul, and the body, in what could be described in popular terms as an out-of-body experience. In his words, looking down upon his own body, he realised: 'This is the state of death. Now I know that I am a soul, that I can exist apart from the body. I shall always believe that, for I have proved it.'[63]

It ought to be pointed out that whereas mainstream Egyptology posits that the pyramids were tombstones or burial chambers, the occult tradition maintains that the pyramids (at least the main pyramids such as the Great Pyramid of Giza) were ancient fanes (temples) of initiation into the Mysteries stretching far back to Atlantean times.[64, 65] The initiation process was conducted, as Brunton recounts in his case, by the guardian-spirits of the Pyramid—High Priests of an ancient Egyptian cult. They conveyed (telepathically, of course) the following message to him: 'Thou hast now learned the great lesson. *Man, whose soul was born out of the Undying can never really die*. Set down this truth in words known to men. Behold!'[66]

Needless to say, the sceptics will mock all this as laughable nonsense and to use their favourite term when all else fails, 'delusion'.

NOTES

1. Manly P. Hall, *The Secret Teaching of All Ages – An Encyclopedic Outline of Masonic, Hermetic, Qabbalistic and Rosicrucian Symbolic Philosophy* (Los Angeles, California: The Philosophical Research Society, Inc. 1988), Preface to the Diamond Jubilee Edition.
2. Nicholas Goodrick-Clarke, *The Occult Roots of Nazism* (New York: New York University Press, 1992). The use of the term 'occult' in the title is correct, but singularly unfortunate. We stress that the term as used in Goodrick-Clarke's book refers to the utter misuse of occultism by the Nazis. Readers should never automatically equate the occult with heinous crimes—see p. 200.
3. *STA*, 'The Fraternity of the Rose Cross', CXXXIX.
4. *Confessio Fraternitatis* (Cincinnati, Ohio: Emperor Norton Books, 2000), as quoted in *STA*, 'Rosicrucian Doctrines and Tenets', CXLII.
5. G. de Purucker, Studies in Occult Philosophy (Pasadena, California: Theosophical University Press, 1973), 637.
6. G. R. Jain, *Cosmology: Old and new* (New Delhi: Bharatiya Jnanpith Publication, 1991), xxv–xxvi.
7. *SD*-II, 'Scientific and Geological Proofs of the Existence of Several Submerged Continents', 794, 797.
8. "Miraculous Powers", Sutras 1–12', in Patanjali, *How To Know God: The Yoga aphorisms of Patanjali,* trans. and commentary by Swami Prabhavananda and Christopher Isherwood (US: Vedanta Press), 1996.
9. Jacobo Grinberg Zylberbaum, 'Human Communication and the Electrophysiological Activity of the Brain', *Journal of Subtle Energies*, 3/3 (1994), 25–43.

10. Stephen M. Phillips, *Extra-sensory Perception of Quarks*, introd. E. Lester Smith FRS (Wheaton, Illinois: Theosophical Publishing House, 1980), 3.
11. *SD*-I, 'Proem', 14–18.
12. Bowen Notes, 8. The Bowen Notes are extracts from the notes of personal teachings given by H.P. Blavatsky to private pupils during the years 1888 to 1891. These are published in a pamphlet entitled *Madame Blavatsky on How to Study Theosophy* (London: Theosophical Publishing House), 1960.
13. Ian Tattersall and Jeffrey H. Schwartz, 'Hominids and Hybrids: The place of Neanderthals in human evolution', *Proceedings of the National Academy of Sciences of the United States of America*, 96/13 (1999), 7117–19.
14. *The Week*, 27 February 2016, 19.
15. Interview by Barry Rohan, *Detroit Free Press*, 25 October 1983; quoted also in *The Eclectic Theosophist* (San Diego, California, May-June 1984), 5.
16. *TSGLOSS*, 140.
17. A paraphrased version of Blavatsky's incisive remark: 'Scalpels and microscopes may solve the mystery of the material parts of *the shell of man*: they can never cut a window into his soul to open the smallest vista on any of the wider horizons of being,' *CW*-VIII, 'The Science of Life', 241.
18. Considerably abridged from *STA*, 'The Life and Writings of Thoth Hermes Trismegistus', XXXIX–XL.
19. Annie Besant, *Seven Great Religions* (Adyar, Madras: Theosophical Publishing House, 1972), 65–6.
20. Better understood in the Sanskrit terms *manvantara* and *pralaya*, meaning cosmic awakening and cosmic slumber. The unsophisticated terms 'Big Bang and 'Big Crunch' in modern cosmogony are faint echoes of this principle of periodic and endless cosmic emergence (awakening) and

cosmic withdrawal (slumber)—see Edi D. Bilimoria, *The Snake and the Rope: Problems in Western science resolved by occult science* (Adyar, Madras: Theosophical Publishing House, 2006), 204.

21. Organization of American States (OAS), *The Incas* (booklet) (Washington, DC, 1975), 3.
22. *STA*, 'American Indian Symbolism', CXCIII–CXCVI.
23. *STA*, op. cit., CXCIII.
24. *STA*, op. cit., CXCIV.
25. Evidence of Inca and pre-Inca spirituality can be traced in: Brother Philip [George Hunt Williamson], *Secret of the Andes: Brotherhood of the Seven Rays* (London: Corgi, 1973); Antón Ponce de León Paiva, *The Wisdom of the Ancient ONE* (California: Bluestar Communications, 1995); Antón Ponce de León Paiva, *In Search of the Wise ONE* (California: Bluestar Communications, 1996); Mark Amaru Pinkham, *The Return of the Serpents of Wisdom* (Kempton, Illinois: Adventures Unlimited Press, 1996).
26. B. F. de Costa, *The Pre-Columbian Discovery of America by the Northmen* (Albany, New York: Joel Munsell's Sons, 1980) <http://tinyurl.com/ya4asfk8> accessed 14 December 2019.
27. Albert Amao Soria, 'The Cosmovision of the Incas', NEXUS (August–September 2017), 48–54, 75.
28. One of many authentic accounts is by D. S. Allan and J. B. Delair, *Cataclysm! Compelling Evidence of a Cosmic Catastrophe in 9500 B.C.* (Rochester Vermont: Bear & Company), 1997. See also: Paul A. LaViolette, *Earth Under Fire* (New York: Starlane), 1997; Genesis of the Cosmos (Rochester, Vermont: Bear & Company), 2004.
29. Brother Philip, op. cit., 13.
30. Serge Raynaud de la Ferrière, 'Message 1: The Coming of the Great Instructor of the World – The Order of Aquarius' <http://tinyurl.com/y8suo25u> accessed 13 June 2018.

31. David Ferriz de Olivares, *Teoría Científica de la Cosmobiología* [Scientific Theory of Cosmobiology] (Peru, Trujillo: Universidad Nacional de Trujillo, 1976).
32. Earlyne Chaney, *Lost Empire of the Gods* (California: Astara Inc. 1994), 137.
33. Grace F. Knoche, *The Mystery Schools* (2nd edn, rev., Pasadena, Californa: Theosophical University Press, 1999), Chapter 6, 'Degrees of Initiation'.
34. G. de Purucker, *Studies in Occult Philosophy* (UK: Theosophical University Press, 1973), 637.
35. ——*Studies in Occult Philosophy*, 638.
36. ——*Studies in Occult Philosophy*, 637.
37. *ibid.*
38. *IU*-I, 'Egyptian Wisdom', 557–8.
39. Paul Brunton, *A Search in Secret Egypt* (London: Rider and Company, 1954), 75.
40. David Frawley, *Gods, Sages, and Kings: Vedic secrets of ancient civilizations* (US: Lotus Press, 2001), 295.
41. *SD*-I, 'The Theogony of the Creative Gods', 424 et seq.
42. Alice Bailey, *A Treatise on White Magic* (2nd rev. edn, London: Lucis Press, 1951), 378.
43. John Gordon, *The Path of Initiation: Spiritual evolution and the restoration of the Western mystery tradition* (Rochester, Vermont: Inner Traditions, 2013), 511.
44. See: Peter Coviello, 'Transcendentalism', in *The Oxford Encyclopedia of American Literature* (New York: Oxford University Press, 2004); 'Transcendentalism', Wikipedia (last modified 10 January 2020) <https://en.wikipedia.org/wiki/Transcendentalism> accessed 16 January 2020. 'New Thought', MSN *Encarta*, Microsoft (retrieved 16 November 2007), accessed 11 May 2020.
45. Aubrey Sherman, *Wizards: The Myths, legends, and lore* (US: Adams Media, 2014).

46. 'New Thought', MSN *Encarta,* Microsoft (retrieved 16 November 2007), accessed 11 May 2020.
47. Arthur Versluis, *American Transcendentalism and Asian Religions* (New York: Oxford University Press, 1993).
48. Horatio Willis Dresser, *A History of the New Thought Movement* (New York: TY Crowell, 1919), 154.
49. 'A Preacher for Trump's America: Joel Osteen and the Prosperity Gospel: Lakewood Church's $60m "smiling pastor" holds up worldly success as proof of God's favour', *Financial Times,* 18 April 2019.
50. 'US Preacher Asks Followers to Help Buy Fourth Private Jet', BBC News, 30 May 2018.
51. Poppy Noor, "Satanic Wombs": the outlandish world of Trump's spiritual adviser', *The Guardian,* 27 January 2020 <https://www.theguardian.com/us-news/2020/jan/27/satanic-pregnancies-trump-spiritual-adviser-paula-white-outlandish-acts> accessed 8 September 2020.
52. Ralph Waldo Emerson, 'Essay IX: The Over-Soul', in *Essays: First Series* (US: BiblioLife, 2009).
53. ——'Essay II: Self-Reliance', in *Essays: First Series* (US: BiblioLife, 2009).
54. ——'Essay IX: The Over-Soul'.
55. ——'Essay I: History', in *Essays: First Series* (US: BiblioLife, 2009).
56. ——'Lecture II: Plato; or, the Philosopher'.
57. ——'The American Scholar: An Oration delivered before the Phi Beta Kappa Society, at Cambridge [Massachusetts], August 31, 1837' <https://emersoncentral.com/ebook/The-American-Scholar.pdf> accessed 17 December 2019.
58. Ralph Waldo Emerson, 'Essay III: Compensation', in *Essays, First Series* (US: BiblioLife, 2009).
59. *The Journals and Miscellaneous Notebooks of Ralph Waldo Emerson*, ed. William H. Gilman, et al., 16 vols (Cambridge: Harvard University Press, 1960, 1982), ix, 211–12.

60. Ralph Waldo Emerson, 'Essay IX: The Over-Soul'.
61. ——'Essay I: History'.
62. Ralph Waldo Emerson, 'Lecture III: Swedenborg; or, The Mystic', in *Representative Men: Seven Lectures*.
63. Paul Brunton, 'A Night Inside the Great Pyramid', in *A Search in Secret Egypt* (London: Rider and Company, 1954), 74.
64. John Gordon, *Land of the Fallen Star Gods* (Orpheus Publishing House), 1997.
65. Willem Witteveen, *The Great Pyramid of Giza: A modern view on ancient knowledge* (Frontier Publishing), 2012.
66. Paul Brunton, *ibid*.

Part Three

The Composition of Man

As microcosm of the universe, man mirrors in his compound nature all elements, forces, and powers of the universe, the macrocosm. Man is essentially a permanent and immortal principle.

Mahatma Gandhi (1869–1948) was well aware of what divinity would mean to a poverty-stricken man when he famously declared: 'To a hungry man, a piece of bread is the face of God.'[1] Of course, he was referring to the starving millions in his homeland, India, who see God's will either as a cause of abundance or deprivation. In contrast, the vast majority of people in affluent countries have more than enough bread, meat, and wine on their dinner tables to render obesity a ubiquitous topic in the news.[2] Coupled with this, we see major portions of society plagued by conflict, violence, and depression. Why? Because there is also a different kind of poverty which is just as oppressive and a lot more insidious. It is spiritual poverty—though the West would hardly acknowledge this as a spiritual crisis.

It would appear that, like the human face, the 'face of God' has two sides to it. The first side concerns material and physical conditions. The second side relates to man's spiritual hunger which is not appeased in the wealthy nations by resorting to atheism or an obsession with technology and wealth, nor with escapism through frivolous entertainment, drugs, sexual indulgence, or alcohol.

Science is marvellously equipped to investigate the physical aspects of man, meaning, of course, his body and brain but it cannot respond adequately to questions concerning the subtler levels of man's mind and consciousness due to its self-limiting mechanistic paradigm. The contemporary mainstream neuroscientific diktat equates thought and consciousness with functional dependence on a machine—the brain. The overall standpoint of mainstream biology and neuroscience has hardly changed during the twenty-first century.

Machines and mechanisms, however, do not run themselves. They require an operator who stands on a level higher than what is being operated. This book accordingly argues that man in his innermost self is an immortal *spiritual* being clothed in a mortal animal body—his functioning mechanism.

The profoundly significant point that emerges from this realisation is the stark error of establishment science in viewing the whole human being as comprising just his physical counterpart—his animal body and personality. Such an attitude means that the inner spiritual being is ignored in favour of its outward mode of expression. In other words, the mechanism is not distinguished from its informing principles.

Scientists have come up with the following composition of the universe: ~68% dark energy, ~27% dark matter, ~5% baryonic matter (i.e., all objects made of normal atomic matter).[3] Therefore, unless everything on Earth, including we humans, somehow manages to exist outside the universe, only some 5% of a human being would be 'normal matter' and the remaining 95% would be invisible to the physical senses.

Is it reasonable, therefore, to postulate that since the greater part of a human being may not be detectable by physical science that it cannot be detected at all? Far from it, as there is a robust case for drawing upon occult science which has studied both the physical and the non-physical aspects of nature and humans in minute detail. Could some 95% of a human being comprise his subtle bodies, namely, his non-physical vestures of consciousness?

Occult Science on the Makeup of Man

Man represents a regular and progressive scale of 'principles' spanning the complete spectrum from spirit to matter. But we must distinguish, carefully, between the inner man and the outer man. From remotest antiquity humanity has had intimations of an internal spiritual entity—a subtle body—'within' the personal physical body.

The Mystery-centres we have discussed in the previous chapter assimilated and taught occult truths of a broadly similar nature. But such truths were formulated according to the language, idiom, symbols and metaphors of their times so it is hardly surprising that the esoteric literature, from archaic antiquity down to the present day, should be written in different languages.

Referring to Figure 1 below, the first column makes the point that man is a unified entity but he displays two fundamental aspects of his being: the inward spiritual and the outward physical. Therefore, we can divide man into two distinct, but interrelated, aspects: the Individuality or Higher Self and the personality or Lower self as seen in the second column. These dual aspects are composed of various principles (shortly described) that in their combinations give the well-known threefold or sevenfold constitution.

The threefold constitution seen in the third column divides man into Spirit, Soul and Body. This classification was adopted by Plato, Saint Paul, and many others. This can be expanded to a sevenfold constitution shown in the fourth column with the terms explained in the adjacent column.

Referring to the sevenfold constitution, the first three principles—Divine Self, Intuition, and Mind constitute the immortal upper triad comprising the Higher Self or Individuality. The last four—Desire, Life principle, Etheric double (Astral body), and Physical body constitute the perishable lower quaternary or perishable personality.

Note, however, that there seems to be little consistency amongst various authors (including Blavatsky) regarding the numbering of the principles. Some work 'downwards', i.e., by assigning 'one' to the spiritual and 'seven' to the physical whilst others work 'upwards', by assigning 'one' to the physical and 'seven' to the spiritual. The convention adopted in this book is strongly recommended because it is logical for numbering to commence *from the point of emanation*—the spiritual.

The role and function of the Individuality and personality are now outlined.

The Permanent Individuality

1. ***Atma*** – The Sanskrit word meaning 'Pure Consciousness' represents the Divine Self, otherwise known as Universal Spirit or Supreme Soul. This is the highest part of man. It is that essential and radical faculty or power which gives every entity its knowledge and sentient consciousness of selfhood. In the human, it is that part of us which is universal in its aspect, rather than individualised. Atma can only contact the lower planes by means of a base or conducting medium and so it is linked with its vehicle, *Buddhi*; as Buddhi is linked with *Manas* and so on down the scale to the physical body.
2. ***Buddhi*** – the Sanskrit word meaning intuition or discernment commonly translated as 'to enlighten', 'to know' and 'to awaken' and therefore 'to understand' or 'to judge'. Note, carefully, that by intuition we mean the faculty of direct knowing, or direct perception, related to, but a higher form of, 'hunches' or 'gut feelings'.

   In one sense, Buddhi may be said to be the germ or nucleus behind Manas, the Mind Principle, but is not the same as reason and still less intellect. Reason, rationality, and intellect are mental faculties and therefore aspects of the mind, whereas the spiritual soul, which implies divine perception and understanding, goes directly to the hidden meaning and therefore transcends rationality—the faculty at work here is in-tuition.

Buddhi in conjunction with Atma constitutes the Human Monad.

3. ***Manas*** – the Sanskrit word meaning 'mind' derived from the verbal root *man,* meaning 'to think' or 'to reflect'.
   Manas springs forth from Buddhi and in conjunction with the Desire principle (see below) is a fundamental pivot point in the human constitution, for it links the Individuality with the personality.
   The Mind Principle is divided into Higher Manas and Lower manas. In other words, mind can either 'rise' and ally itself with Buddhi and bring forth the spiritual nature or it can 'descend' and entangle with *Kama,* the Desire principle and thereby inflame the animal nature. But this does not mean that man has two minds working simultaneously within him! Nor does it imply a literal association with the two hemispheres of the brain. It *means that the same mind is working in different ways and at different levels. This duality in the functioning of mind is absolutely crucial to our understanding of how man functions.*

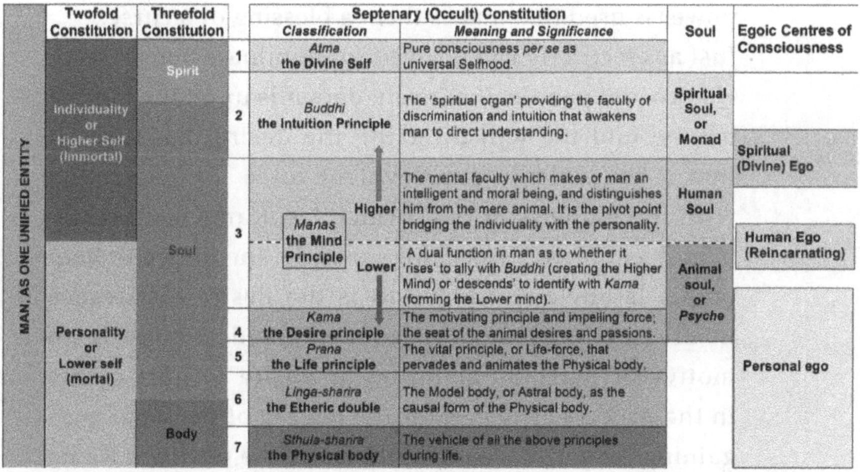

MAN, AS ONE UNIFIED ENTITY	Twofold Constitution	Threefold Constitution		Septenary (Occult) Constitution			Soul	Egoic Centres of Consciousness
				Classification		Meaning and Significance		
	Individuality or Higher Self (Immortal)	Spirit	1	*Atma* the Divine Self		Pure consciousness *per se* as universal Selfhood.		
		Soul	2	*Buddhi* the Intuition Principle		The 'spiritual organ' providing the faculty of discrimination and intuition that awakens man to direct understanding.	Spiritual Soul, or Monad	Spiritual (Divine) Ego
			3	*Manas* the Mind Principle	Higher	The mental faculty which makes of man an intelligent and moral being, and distinguishes him from the mere animal. It is the pivot point bridging the Individuality with the personality.	Human Soul	Human Ego (Reincarnating)
	Personality or Lower self (mortal)				Lower	A dual function in man as to whether it 'rises' to ally with *Buddhi* (creating the Higher Mind) or 'descends' to identify with *Kama* (forming the Lower mind).	Animal soul, or *Psyche*	Personal ego
			4	*Kama* the Desire principle		The motivating principle and impelling force; the seat of the animal desires and passions.		
			5	*Prana* the Life principle		The vital principle, or Life-force, that pervades and animates the Physical body.		
			6	*Linga-sharira* the Etheric double		The Model body, or Astral body, as the causal form of the Physical body.		
		Body	7	*Sthula-sharira* the Physical body		The vehicle of all the above principles during life.		

Figure 1: The Metaphysical and Physical Composition of Man

The Temporary Personality

4. ***Kama*** – the Sanskrit word meaning 'desire' derived from the verbal root *kam*. (This is not to be confused with karma, the law of cause and effect.)

   Kama is the driving, or impelling force. In the human constitution this force is 'the seat of animal desires and passions'.[4] It is the centre from which operates the living electrical impulses, desires, and aspirations. This fourth principle, in conjunction with Lower manas, is the centre of man. However, Kama as the motivating principle without *Prana* (as the motor principle) would be inert. Just as the wind fans the flames, so as the vital breath Prana makes all desires vital and living.

   There seems to be a curious notion amongst some self-righteous people or in sanctimonious spiritual groups that desires are basically evil. This is nonsense. In itself, Kama is neither 'good', nor 'evil' but like any and every force and power in nature is completely neutral and 'colourless'. It is the manner in which such force and power is used that makes them a blessing or a curse.

   Just as electricity can be used to illuminate houses or to electrocute people, it entirely depends upon the motive behind and the will directing the desire that turns it into a benevolent or malevolent force. In association with the higher spiritual principles, Kama is a greatly beneficent factor, since, for example, the desire to help others is obviously noble as is the desire to advance in life and acquire knowledge—provided there is no motive of personal ambition or desire to hurt others in the process. The greater the degree of personal end-gaining for selfish reasons, the more the purity of Kama is polluted.

5. ***Prana*** – the Sanskrit word meaning 'life-principle', the 'psycho-electrical field, or veil manifesting in the individual as vitality'.
   Other similar terms would be life-force, vital breath or vital current. It is the *chi* of the Chinese and the *fu* of the Japanese. Prana is universally pervasive so that the physical body is pervaded by it as is the Earth. It should be noted that a similar term is *jiva* meaning 'to be alive'. It is necessary only to the mortal elements of man—the physical body, etheric double, desire principle and to all mental functions which operate through the physical brain, namely, the lower mind.
   There are a number of Pranas each having its own name and function. One system mentions three Pranas; five, seven, twelve and even thirteen types are identified elsewhere.[5] These are different aspects of the one life-principle. At the moment of physical death the 'life-atoms' of Prana immediately return to the pranic reservoirs of the planet.
6. ***Linga-sharira*** – the etheric double, model body (astral body), the Sanskrit word *linga,* meaning 'model' and *sharira* meaning 'form' or 'body'. The term signifies impermanence. It is the *doppelgänger* of the Germans and the *eidolon* of the Greeks.
   The term 'etheric' implies something more ethereal than the physical body. It acts as the template around which the physical body is built. It radiates the energies that enable the physical body to be built. Although impermanent, it is not subject to constant change like the physical body, but remains relatively stable during a lifetime. It is of an electro-magnetic nature and remains in close proximity to the physical body.
   The etheric double precedes the physical body but outlasts the death of the latter, remaining in the astral realms and

gradually fading out and dying. The misty or grey looking wraiths that sensitive persons sometimes see hanging around graveyards are the *Linga-sharira* of the recently departed.

7. ***Sthula-sharira*** – the physical body, the Sanskrit word *sthūla,* meaning 'coarse' or 'gross' and *sharira* meaning 'form which disintegrates'.

   Man's physical body serves as the vehicle of all his other principles and aspects during life. In another sense, 'body' is a collective and generalising term to denote the lowest constituent of man. Strictly speaking, the physical body is not really a 'principle' since it acts like a shelter, or as a garment and carrier, of the real man. So, the loss of his body in no way implies the death of the man any more than we would physically die, should we lose our house or our clothing.

Mysteries of the Soul

'Soul' is a very broad term with wide interpretations but is essentially the intermediary vehicle between the spirit which is deathless and the body which is mortal.

Transformers are the necessary intermediaries for stepping down the high voltage electricity generated at power stations to domestic consumers. Soul is the *generic* term for the transformer stations that convey the divine 'current' from the generating station (Atma) to domestic consumers (the physical body). In electrical distribution systems this cannot be achieved by just one transformer. Several are needed for a graded 'step down' from the highest voltage to the lowest. So, the soul functions at different levels to 'step down' the 'spiritual voltage' of Atma to the lower level required by physical man. As shown in the sixth column of Figure 1, there are three 'transformer stations': (*a*) the *Spiritual Soul;* (*b*) the *Human Soul;* (*c*) and the *Animal soul.*

Nothing remains of the third level after physical death. Of the second, only its divine essence, *if left unsoiled* survives while the first, in addition to being immortal, becomes *consciously* divine by virtue of assimilating higher Manas [mind].[6]

The principal message is that the dominance of the physical over the spiritual nature (and in many cases the utter stifling of the spiritual by the physical) is the condition in which many people remain throughout the course of their lives. This is understandable because the physical nature is more immediate and accessible than the spiritual nature. But this need not be so. For those who will to discover the inner world and consciously pursue the uphill path, that 'light upon light' shines ever more brightly so as to fulfil the real purpose of their lives.

The physical aspect of man, like the physical world he inhabits, is, relatively speaking, of minor consequence in this boundless universe. Spiritually, man has endless opportunities to attain divine knowledge and wisdom in the realm beyond.

Profound Mysteries of the Three Selves

There are three vehicles of consciousness which are known in the Mystery Teachings as man's 'Three Selves': the Auric Egg, the Causal, or Karmic Body, and the Mortal vesture. Before outlining their role it is important to explain the significance of the Egoic centres of consciousness in man.

The penultimate and final columns in Figure 1 above show the distinction between the soul and egoic centres of consciousness. In simple terms, 'soul', is essentially a transforming principle functioning on three main levels as explained above. However, in order to actuate as a vehicle of consciousness in man, soul has to engage with the Mind Principle, which conveys to man his consciousness of 'Self'. In esoteric philosophy the Ego is that consciousness in man of 'I am-ness'; or the feeling of 'I am such-and-such' at any level—it does not convey the popular notion of exaggerated self-opinion.

Thus as seen in Figure 1, the Divine Ego is the union of the Spiritual Soul with the Higher Mind; the Higher, or Inner

Ego also referred to as the Reincarnating Ego is the Mind Principle *per se*; and the Personal, or Lower ego the conjunction of the Lower mind and mortal vestures, also referred to as the 'false personality', being the bundle of finite and transitory experiences accumulated during mortal life.

We now outline how the three Egoic centres function through their corresponding vehicles of consciousness—the 'three selves'.

The Auric Egg

This 'envelope' is the source and basis of all other vehicles of consciousness that form the human seven-fold constitution from Atma to the physical body. It has an egg-shaped appearance but is not the same as the human aura.

Its 'substance' is drawn from the universally diffused, primordial and pure substance known as *Akasha*, and for this reason the Auric Egg reflects all the thoughts, words, and deeds of man and endures throughout his cycles of reincarnation. As Blavatsky teaches, the Auric Egg is:

1. The preserver of every karmic record (karma being the universal retributive law of cause and effect).
2. The storehouse of all the good and bad powers of man. This aura is the mirror in which 'sensitives' and clairvoyants sense and perceive the real man and see him as *he is*, not as he appears.
3. That which furnishes man with his astral form around which the physical entity models itself, first as a foetus, then as a child and adult, the astral growing apace with the human being.[7]

The Causal Body

This is the primary vehicle of the Spiritual Ego. The Causal Body is the union of *Buddhi* with *Manas*. The Causal Body is also referred to as the Karmic Body.[8] And also why, strictly speaking, it is the primary vehicle of the Spiritual Ego and Higher Ego.

It is difficult to say much more about this body since, in the main, humanity is not yet capable of appreciating the inner or higher worlds. What may be said is that this vehicle of the Higher Ego is without form or quality. However, on its own plane, and when perceived by a highly advanced person, it does have a form. But the form has no sensory apparatus, since the Higher Ego is not time bound by past-present-future events but operates in the ever-present.

The Mortal Vestures – the Outer Garment

Man's most material vehicle capable of supporting his mental function—his 'temple of the spirit', his physical body, is the principal vehicle of the lower mind which feeds on the images furnished by the senses and co-ordinated by the brain.

The physical body and the Lower mind live and grow in union, the latter storing impressions as memory, the content of which is totally dependent upon the perception of the physical world accessed through its material vehicle. It must be admitted that for the vast majority of people this is their only waking reality for, despite all the talk of other worlds and higher dimensions and planes, few people can claim to have had actual and direct experience of them. However, the dream state, especially lucid dreaming, which is common to a large section of the population, strongly alludes to superphysical realms.

The Body of Illusion

The word *Mayavi-rupa* means the 'illusory body' (*Maya* is illusion and *rupa* is form). This is a higher astral-mental form. Similar terms in German and French are *doppelgänger* and *perispirit*.

It is important to note that the Mayavi-rupa pertains to the living person. It is dual in its function, being 'the vehicle both of thought and of the animal passions and desires, drawing at one and the same time from the lowest terrestrial manas (mind) and kama, the element of desire.'[9]

First-hand accounts of a sensed presence abound in legends and literature, both popular and esoteric, all over the world. One experience was narrated by Sir Ernest Shackleton (1874–1922), the Irish explorer who led three expeditions to the Antarctic. His book *South: The Endurance Expedition* described his belief that an incorporeal being joined him and two others during the final leg of their journey. Shackleton wrote, 'during that long and racking march of thirty-six hours over the unnamed mountains and glaciers of South Georgia, it seemed to me often that we were four, not three.'[10] His admission resulted in other survivors of extreme hardship coming forward and sharing comparable experiences.

Opening Heaven's Door is a recent book providing examples of how survivors from life- threatening situations such as terrorist attacks, drowning, shipwrecks and air crashes were guided to safety by a mysterious presence that was sometimes visible, at other times unseen, or else audible, or sensed and, at times, even bossing, and prodding.[11] In all such cases, when its task is completed the presence disappears without any trace.

Mainstream scientists, of course, look to materialistic arguments and mechanistic theories to explain away such phenomena. For example, rather than acknowledge any kind of presence or unusual experience, they contend that soldiers in the trenches suffered from sleep deprivation; polar explorers were deluded by sensory deprivation due to the white landscape; shipwreck victims hallucinated from sunstroke and dehydration and mountaineers suffered from high altitude lack of oxygen and cold.

In many cases such appearances can be ascribed to the Mayavi-rupa since it is a *vehicle of projection* during earthly life. The circumstances are numerous such as the appearance to a loved one of a dying person, or even a recently deceased person, or by the intervention of a higher being who intercedes to guide or rescue an individual. They cannot all be explained away under the usual labels of 'delusion' and 'hallucination'. Science needs to understand that the higher states of consciousness, and associated spiritual experiences, cannot be replicated to order at the behest of the scientist under what is fashionably known as 'laboratory conditions'.

These experiences can happen when a person is touched by the 'hand of grace' which is drawn only to those who have managed to hold the ego (i.e., self-centredness) in complete abeyance through simplicity and holiness of living or the state of self-forgetfulness common to scientists and artists during moments of peak inspiration.

Suspension of the personal ego can also occur in extreme survival conditions that result in sensory deprivation, rendering a person more receptive to higher and subtler influences.

The Mayavi-rupa can also be directed by a living person towards a relative or loved one or for the purposes of genuine scientific research. 'Lucid dreaming' and 'out-of-body experiences' are the modern terms used for this aspect of projecting the 'thought body' and this can happen spontaneously or, for advanced persons, intentionally, after suitable training.

Thus, at the highest stage of development, an adept can project the Mayavi-rupa at will and in full consciousness. He endows the form with as much of his own mind and consciousness as is required and it is this ability which accounts for the stories we hear of adepts being seen in two places simultaneously.

During the sixteenth, seventeenth, and eighteenth centuries a considerable number of alchemical adepts journeyed throughout Europe, appearing and disappearing apparently at will. According to popular tradition, these adepts were immortal and kept themselves alive by means of the mysterious medicine that was one of the goals of alchemical aspiration. It is asserted that some lived for hundreds of years, taking no food except this elixir, a few drops of which would preserve their youth for a long period of time. That such mysterious men did exist there can be little doubt as their presence is attested by scores of reliable witnesses. One such example is that of the illustrious German alchemist and philosopher Comte de St.-Germain (*circa* 1691/1712–1784) of whom it is recorded that he had the astonishing ability of both appearing in, and departing from, his own apartment, and those of friends, without using the door.[12]

Similar such powers have been attributed to the Bulgarian sage Peter Deunov. A notable incident was when he vanished from a locked room, appeared outside the door knocking on it and re-materialised inside the room—an episode that was repeated three times.[13] There are other accounts where Deunov rescued disciples when in danger and asked them about this afterwards—it is as if they were in his field, or aura. The rationale behind this can be gleaned from his book *Divine Providence*.[14]

It cannot be stressed sufficiently that all such displays of extraordinary powers by highly advanced beings did not happen only in past centuries (hence relegated to mediæval folk-lore by diehard sceptics) but occur also in relatively modern times. Nor have they anything to do with glamour-seeking circus acts but are a wake-up call to 'spiritual sleep-walkers' and rank disbelievers regarding the latent and truly infinite powers in man.

Death – the Door to Eternity

We have to ask the question: 'what exactly do we mean by death?' In occultism, the meaning is not quite as straightforward as the everyday sense of the term. And is there just one death? In fact, there are two deaths, which logically means that there have to be two births. The first death is obvious but the second one needs explaining.

The transition of man's consciousness after physical death and towards subsequent rebirth is now described in three principal and well-defined phase transitions: (*a*) from physical life to astral life; (*b*) from astral life to spiritual life; and (*c*) from spiritual life to a new terrestrial life. The term 'astral' is used as the best available term in English to refer to the psychic realms that encompass the broad, intermediate planes ranging from the physical to the spiritual.

From the standpoint of materialism physical death means the cessation of the functions of the brain, heart, and other bodily organs. This is obviously so but there is more to consider. The popular idea that loss of brain function means the extinction of consciousness is not upheld nowadays by enlightened scientists and psychologists and certainly not in occultism, which teaches that physical death is transition.

In the course of dying and immediately afterwards, the lower sentient consciousness of the brain and physical senses is extinguished. However, the higher consciousness survives forever because its root lies in eternity. Memory is therefore dislodged from the physical brain. Furthermore, the *Mayavi-rupa* (the 'illusory body') may appear to loved ones as a materialisation of the newly departed person during this early transition stage.

Soon after physical death, a review of the past life occurs in *Kama-loka,* also known as Amenti in Egyptian mythology, Hades in Greek mythology and purgatory in Christian theology. All these terms signify the realm of the dead, or rather, a stratified intermediate state of consciousness in which the soul must undergo purification so as to achieve the holiness necessary to enjoy the consciousness of heaven.

Purgatory may be likened to a school examination room. Those who miss the mark are demoted to be disciplined by more hard study and those who pass move 'upwards' to the next higher class. It is always the case that the punishment fits the offence so, fittingly, as depicted in the *Purgatorio* division of Dante's epic, narrative poem *Divine Comedy,* the proud are forced to bend double under the weight of heavy boulders and the gluttons have to chase in vain around a tree laden with delicious fruit.

It must be stressed that the life review here is not in any way concerned with the personal details of the former life but solely with its spiritual content in order that those spiritual elements can be distilled and assimilated into the Higher Self. The profound life review is experienced from many angles. This includes, significantly, the 'inside out' experience of being on the receiving end of causes enacted during life (for example, a person who deliberately causes hurt to a loved one during life will experience the effects of that hurt upon himself). The golden rule of treating others as we ourselves would wish to be treated, a maxim found in practically all religions and cultures, is an inexorable universal law.

With the ending of this review and its consequent striking of the keynote for the next incarnation, we experience the unconsciousness of deep, dreamless sleep.

The Second Death

The life review is followed by what is known in occultism as a 'death struggle' which implies conscious suffering and pain but is the natural separation of the Higher Ego from the personal elements in the man's experience. The struggle is between the Higher Manas and the Lower manas, the former in association with Buddhi and the latter identifying with kama or desire.

Then follows a 'gestation' period in order to assimilate the 'spiritual aroma' of the past life into the Higher Egoic entity. Unless there are unusual circumstances in the former earthly life (such as a premature death by violence, suicide or other means such as accident), the gestation period for the average person may last for many years (in terms of Earth time) after which consciousness returns.

The separation of the immortal parts from the lower portions is profoundly significant and well symbolised by the butterfly emerging from its chrysalis. What this means is that all burdensome encumbrances such as cravings for material possessions, sensual cravings, outworn intellectual concepts and religious encrustations must be sloughed off.

Divested of the impediment of his last mortal remnants, man is reborn as his true immortal Self on the higher plane known as *Devachan* until the hour for a new physical incarnation arrives under the divine law of harmony and adjustment—karma.

Consciousness in the 'Kingdom of the Gods'

Devachan is a compound Sanskrit-Tibetan term deriving from *deva* 'a "god" ' and *chan* 'land', 'region'. It is known by various other names like *Sukhavati* (Sanskrit: literally 'land of bliss') in Mahayana Buddhism, *Elysium* of the Greeks, *Sekhet-Aaru* of the Egyptians, and corresponds with *Paradiso* in Dante's Divine Comedy. Devachan bears a relation to, but is not identical with, the heaven of orthodox Christianity, which regards the heavenly state as permanent and eternal (a good example of how the exoteric adaptation of religion distorts the esoteric meaning). Devachan is a temporary state between two Earth lives into which the Higher Ego enters after the second death when divested of its lower aspects.

Devachanic life and consciousness can, in one sense, be seen as a counterbalance to life on Earth.

Devachan (like Kama-loka) is not a geographical place but a spiritual state of consciousness where time has a very different connotation from the everyday world. Devachan is emphatically not a monotonous repetition of some former pleasing experience, but the 'workshop' to which an individual's 'rough diamonds'—good thoughts, words, and deeds—are taken for fashioning after being extracted via much suffering and toil from the 'mine' of the personal life. It involves complete engrossment in the bliss of all former personal earthly affections, preferences, and thoughts. All the unfulfilled spiritual and higher intellectual possibilities of the former life are constantly reviewed and improved in the imagination.

The central role of Devachan is the fulfillment of all *unfulfilled* spiritual yearnings of the past incarnation. Being a spiritual state, no frustrated earthly or material ambitions, nor any personal desires or fantasies, can have any relation to Devachan. These are all associated with the Kama-rupa, namely, the subjective form created by means of the thoughts and desires of a person during life projected as a form into the astral world after the death of the physical body, which, like the mortal coil earlier, has been sloughed off to disintegrate over time.

Like Kama-loka, Devachan is stratified in levels to match the spiritual status of the Higher Ego. Interestingly, the Eddas (two thirteenth-century AD Icelandic manuscripts that are the main sources of Norse mythology) describe Valhalla as a majestic, enormous hall ruled over by the god Odin and containing five hundred 'doors'—each door corresponding to a different Valhallic condition, i.e., state of consciousness.[15] Just as the punishment fits the crime in purgatory, so the reward fits the meritorious deed in paradise. The artist, scientist, poet, or philosopher will rise to their appropriate 'grade' to reap the fruits of good deeds sown on Earth and fulfil yet higher aspirations.

The phenomenon of genius can, arguably, be explained by innumerable incarnations and their associated devachanic experiences, ever refining the faculties and renewing the aspirations of former lives.

To whatever level the 'devachanee' may 'rise', there is an absolute oblivion of all that gave pain or sorrow in the past incarnation and even oblivion of the fact that such things as pain or sorrow exist at all. The devachanee lives his intermediate cycle between two incarnations surrounded by everything noble that he had vainly aspired to and in the companionship of everyone he loved whilst on Earth. He has reached the fulfilment of all his soul-yearnings. And thus, for long centuries, he may live an existence of unalloyed happiness, which is the reward for his sufferings during Earth-life and the fruit of worthy actions.

That which is conscious in Devachan is the spark of consciousness that preserves in the Spiritual Ego (Buddhi and Higher Manas) the idea of the personal 'I' of the last incarnation. This lasts as a separate distinct recollection only throughout the devachanic period, after which time it is added to the pool of innumerable incarnations of the Reincarnating Ego. Immortality is one unbroken stream of consciousness. Hence the personal consciousness can hardly be expected to last longer than the personality itself. Such consciousness survives only throughout Devachan, after which it is reabsorbed, first, in the individual, and then in the universal consciousness.

Who Enters Devachan?

Occultism teaches that not all who physically die enter Devachan. For example, children who die young (usually before the age of seven), get reincarnated quickly as they have not yet gathered the experiences that will reap a blissful condition in Devachan. So, also heinous criminals who obviously, having reaped practically no spiritual harvest in life, are therefore reborn swiftly. Other exceptions are some of those who die prematurely and who may be reborn immediately, those who commit suicide, and victims of murder, or accident.

Further exceptions are advanced souls who defer Devachan and choose to be reincarnated quickly in order to serve humanity in physical life. At the highest level of advancement are initiates and adepts who have transcended the illusory veils of the mind and therefore have no need of Devachan.

From the above, it may correctly be inferred that death is indeed a promotion in the classroom of life! Other than the few cases just cited, the vast majority of mankind are indeed 'promoted' to the higher and purer life of Devachan until such time as each soul is prompted by karmic law to commence a fresh chapter and thus incarnates into a new classroom in the stern school of life.

The Nature of Time

It is natural to enquire about the length of time spent in Devachan. But, as previously stressed, the clock time measured on Earth is very different from the subjective sense of time. It is commonly known that what seemed to last for a long time in a dream, in fact took only a few seconds, as shown by measurements of the rapid eye movement (REM) of the dreamer, indicating the onset and interval of the dream state.

The overriding principle is that the period in Devachan is *proportionate to the unexhausted spiritual impulses* originating during Earth life. However, the time in Devachan is enormously greater than time according to Earth life. Moreover, such time varies widely with different human beings. Since the devachanic experience is meant to be an elaboration of what the human being has learnt during his physical existence, to unfold it freely, to make it suitable to a new life, it is obvious that the time spent in Devachan is proportional to the 'devachanic store' of the individual. A man who has immersed himself in little more than the life of the five senses would dwell only a short time in this abode to 'process' the spiritual content of his earthly experiences. Conversely, a man who has collected rich experiences obviously has to process a lot more and hence, has a longer stay in Devachan.

Depending on the degree of spirituality and the merit or demerit of the last incarnation, that interval can last a few minutes, one year or a million years. The average time (from the standpoint of Earth-time and not to be taken literally), as we are informed, would be around fifteen hundred years for the majority of those who wish for a release and for an enjoyment of bliss.

The Question of Immortality

Death and immortality are the two opposing polarities of time: the former approaching the infinitely small, the latter the infinitely large, but both in the realm of time. Therefore, immortality is not exactly the same as eternity. Immortality has a beginning but no ending. Eternity has no beginning and no end.

Immortality means a continuous and unending existence or being. But there is nothing in nature that remains static and does not evolve over time. What varies is the periodicity, or time periods of change, not the fact of change. For example, compared to the mayflies (also called 'one-day insects') who live for twenty-four hours and some species that die within a few hours, the average life span of the modern American is seventy-nine years.

What is indeed immortal is the uninterrupted stream of consciousness, albeit undergoing continuous and unceasing changes of states in its realisations of itself throughout endless duration.

The Sanskrit word *manvantara* means a period of manifestation pertaining to the various life-cycles, terrestrial, planetary, solar and beyond. *Pralaya* means a period of rest and dissolution. They may be thought of in the most general sense as 'world periods', but readers should refrain from making direct comparisons with the postulated age of the universe according to physical cosmology. (For interest, the length of a planetary manvantara, also known as the Day of Brahma, is 4,320,000,000 years.) From the standpoint of man, then, immortality refers to the immense period of time spanning a world period, after which the divine entities thus passing out enter into still higher realms to reappear at the end of pralaya at the dawn of the succeeding manvantara.

The Issue of Reward and Punishment

The Higher Ego in the devachanic state is not omniscient and is not aware of what is happening in the physical world that it has left behind. The karma of evil deeds is temporarily suspended, and only the karma of benevolent thoughts and deeds are carried into Devachan. Suffering is reserved for the subsequent Earth life, not in the post-mortem states, because *the causes generated on any plane have to meet with their consequences on the same plane*. What is sown in the physical world must be reaped in the physical world. In other words, causes generated on Earth can exhaust in no other state but on Earth. There are exceptions, however.

For example, a case of suicide where an individual retains a degree of consciousness and remains for a while in an unhappy limbo state for having been deliberately responsible for taking their own life. However, such a person's post-mortem state would depend entirely on their motive for ending their physical life. There is an absolute world of difference in motive between a person who takes their life because of, say, unbearable physical pain, or to avoid terrorist-interrogation under torture in prison, and another who decides to do so to escape the course of justice for some criminal act.

Suffering may also be the lot of a person who dies in the clutches of some abnormally strong unsatisfied ambition or an uncontrollable passion for the physical satisfactions of food, drink, sex, or material possessions. Extreme sorrow also inhibits a departed soul from going ahead towards greater peace and freedom. Other than such exceptional cases where an individual brings suffering upon himself by excessive attachment to gross physical appetites or excessive grief, there is no post-mortem punishment for evil or misguided conduct when in the body, nor is there any experience corresponding to *eternal* damnation or the traditional hell. But, as always, there is a major exception to the common rule: it is the state known as *Avitchi* and this is touched upon later.

The Third Transition: A New Life on Earth

Reincarnation is the cyclical process whereby the Ego in Devachan takes on entirely fresh mortal vestures for a new life on Earth. Strictly speaking, then, rebirth refers to the physical event as a final outcome of the cyclic process of reincarnation. However, only a fraction of the spiritual nature 'descends into', and clothes itself in material existence for an individual life on Earth.

The devachanic experience goes through stages analogous to Earth life—the post- mortem equivalent of physical growth and death: the first flutter of sentient life, the attainment of 'prime' growth, the progressive exhaustion of force passing into semi-consciousness, gradual oblivion and lethargy, total oblivion and then—not death, but rebirth into another personality on Earth.

In daily life we are naturally drawn to revisiting familiar experiences, say a favourite restaurant or a memorable holiday. Just so, what draws the devachanee back into rebirth is the force of *Tanha,* a Pali word familiar in Buddhism, meaning the 'thirst' for material life, the desire to live and cling to earthly life. But what is this thirst based on? *Trishna* is a Sanskrit word, which also means 'thirst', but with the added qualification of thirst for things, and familiar scenes, which the human ego formerly knew from past experience and which it wills and desires to know again.

Obvious attractors are desires unfulfilled, love unrequited, duties left incomplete and yearnings unfulfilled. All thoughts and actions sown must of necessity reap their fruits on the Earth plane where they were sown. This thirst for the familiar, material life which brings the Reincarnating Ego along the downward arc back to Earth-life is the strongest individual cause for reincarnation.

Reincarnation is a highly complex subject with numerous interfaces: pre-existence, re- embodiment, metempsychosis, and transmigration—all related, but not the same. However, no comprehensive account of reincarnation can be given without a full treatment of the Law of Karma.

At the solemn moment of death each man sees in a flash the whole of his life marshalled before him and is shown the whole chain of causes that have been at work during his life. Similarly at the moment of rebirth, the Higher Ego has a prospective vision of the life which awaits him and realises all the causes that have led to it.

All the vices, desires and, especially, the passions of the preceding incarnation become, through definite Laws of Affinity and Transference, the germs of the future potentialities in the animal soul (Kama) with its dependence on the astral double.

The Higher Ego—the *Christos* principle in each man—remains always the same. It is only its unruly child, the personality, that changes from one incarnation to the next. And it is the Ego's karma that guides and dictates the moral traits of its former child (the old personality).

It must not be imagined that parents provide a soul to the incoming reincarnating entity. Human parents bequeath only the appropriate physical vehicle and psychic vesture to the reincarnating soul, which must then work out its own karma through its new, karmically conditioned psycho-physical body. For example, an advanced 'musical soul' would naturally be drawn to parents who can provide a highly sensitive nervous system.

The popular idea about the human soul reincarnating as, say, an animal or even a stone, as a punitive measure for evils committed in an earlier existence, and then having to work its way up through the kingdoms of nature by way of moral rectitude, is to misunderstand both the nature of the human soul and reincarnation. Deeply rooted in *orthodox* Brahmanical teachings, this idea is another gross distortion of the original meanings of metempsychosis and transmigration of souls. *Once a human, always a human* is the law and other than in the rarest of abnormal circumstances, which need not be entered into here, it is virtually inviolable.

Spectres and Apparitions

What do we make of the stories that abound in the legends and fables of all cultures and religions about what are generally known by names like 'spooks' or 'ghosts'? How do such apparitions occur, and do they have any consciousness to speak of?

After the second death, whereupon man is 'born again' and his spirit has 'ascended' to Devachan, there can obviously be no return to Earth, since there is then no connecting link. Released of its divine inhabitant the Higher Ego, the Kama-rupa, becomes a psychic corpse known in occult literature as a 'shell'. Just as a cast-off overcoat may retain the shape and the odour of its former wearer for a time, so the discarded vesture (shell) of the Higher Ego retains some of the characteristics of the deceased personality.

Left to itself, the discarded Kama-rupa of the late personality—the shell—disintegrates when its residual psychic energies are dissipated and its constituents recycled back to the psychic planes of the universe. However, when the discarnate man is still strongly earthbound and unable to loosen his earthly ties to persons, places, and possessions, the shell can hover in the terrestrial atmosphere and sometimes appear as an apparition. Séance room phenomena are mainly due to the artificial activation of the shell by the vitality of the medium or sitters.

The duration of the Kama-rupa, and subsequently the shell, depends entirely on the sensual energies of the man during life, in other words, the extent to which his thoughts have energised and fuelled his desire nature, which, if not fully satiated or exhausted during physical life will have to expend its pent-up energies after death.

If the man has turned his thoughts towards the higher life then the shell may exist for a short time before disintegrating. But, if forcibly drawn back into the terrestrial sphere the 'spook' may prevail for a period greatly exceeding its natural existence in Kama-loka. There are two ways in which this can happen.

If during earthly life the man harboured strong passions or died with unfulfilled lustful appetites or with thoughts of murder, violence or revenge, then the spook, comprising those pent-up psychic energies could be attracted earthwards by the atmosphere of debauched places and the aura of dissolute persons or, worst of all, by abominable necromantic practices. Once the Kama-rupa has 'learnt' the way back to living human bodies (i.e., established a magnetic affinity with a person), it becomes a vampire, feeding on the vitality of those who invited its company.

In India these eidolons are called *Pisachas* and are much dreaded, as also in other cultures. It is precisely for this reason that capital punishment, considered from the occult standpoint, is not to be recommended. A murderer in his body serving his life sentence in prison is a lesser threat to humanity than one forcibly thrown out of his body by execution with all his criminal propensities in full flood—such only serve to further poison the psychic atmosphere.

The other case is that of a perfectly good man who has suffered a sudden violent death in the prime of life. Because such a man has not lived his allotted life span, the Kama-rupa will be charged with unspent psychic energies and affinities, compelling him to complete unfinished business. He may then be drawn into the terrestrial atmosphere by the excessive grief of friends or loved ones. Such a spook, although startling to the beholder, is harmless and benign.

There have been many reports of soldiers killed in World War II whose apparitions later appeared in their homes. Their surviving colleagues have reported that when mortally wounded, many a young soldier cried out to his mother at the moment of death. It is not difficult to envisage that a strong emotionally-charged final thought of a dying youth towards his home and mother, reinforced by his mother's own constant loving thoughts towards her son on the battlefield, would result in a materialisation in the family home.

Occult science counsels that overwhelming and extended grief, beyond the natural course of bereavement, is selfish because it holds back the departed from their onward journey. The real tragedy is for those left behind in their personal bereavement, not for the newly departed who have been freed of all sorrow and pain, albeit temporarily.

Avitchi and the Eighth Sphere

Avitchi is a Sanskrit word meaning 'waveless', 'without happiness', or 'without repose', implying total stagnation of life with no happiness and continuous agony. Avitchi may be regarded as the counterpart, or 'shadow', of Devachan. It afflicts all those who have slipped down into the mire of unredeemable sin and bestiality. When that happens, the stream of personalities eventually comes to an end because the Higher (immortal) Ego will have lost all influence over its lower selves and the connection is severed.

The Higher Ego is also affected by such a breakdown. Having failed to gain the experience of earthly life that its personalities were supposed to supply, it too becomes a failure in its own realm and falls into a dreamless sleep awaiting the next planetary cycle to re-evolve its humanity, this time in full consciousness through all the forms of the lower kingdoms of nature.[16]

Does this imply that there are states even lower than the physical and even more suffused in selfishness and materiality? The occult doctrines do not shirk from pointing out that this is indeed so and carry a heavy warning that humanity must have no truck with such states which belong to a distant past. There can be breakouts from such subterranean vaults of consciousness whereby individuals or groups have inflicted unspeakable brutality upon humanity in ways psychiatry and psychology are at a complete loss to explain.[17] The crimes perpetrated by the Nazis upon humanity, intensified in their horror by the mental element of strong premeditation, meticulous planning, and ruthless organisation, are just one of many recent examples.

But Avitchi is not necessarily a post-mortem state only, or between two births. It can also exist on Earth. Avitchi is not a place of eternal damnation or 'punishment' in the orthodox Christian sense but a generalised term for places, namely, *states of consciousness* of extreme evil *and the final extinction of the evil entity itself, ground over in nature's laboratory*. This is also known as the *Eighth Sphere* or *Planet of Death*.

There is a state of psycho-mental degeneration, more advanced even than Avitchi, which, in the esoteric philosophy, is called the habitat of 'lost souls'. Where there is the glimmer of re-ensoulment in Avitchi by way of the umbilical connection with the monad (the conjunction of the Divine Self, Atma, and Buddhi.), such a possibility vanishes in the Eighth Sphere. We are informed that whereas Avitchi is exclusively a state or condition, the Planet of Death is an actual globe.

Just as there are grades of 'heavens', Avitchi also has many grades where those who must find an outlet for their burning evil passions sink to the appropriate level. Here, the punishment fits the offence but much more severely because of the greater severity of the crimes, for example, as depicted in the *Inferno* section of Dante's *Divine Comedy,* being imprisoned in an icy lake in perpetuity or wandering forever lost in a dark wood confronted by savage beasts reflecting the savage thoughts and deeds put out as causes during life.

Antahkarana – Connecting the Personal and the Divine

The essential meaning behind *Antahkarana* is the intermediate vehicle of consciousness functioning as a narrow *bridge* between the lower and the Higher Manas. Some Australian Aboriginal tribes speak of a similar 'bridge' between the lower self and the spirit (in popular terms) which they call 'the Rainbow Snake'.[18]

The Antahkarana conveys to the Higher Ego all those personal thoughts of its lower counterpart so they can be assimilated and be made immortal with it; but only those thoughts and aspirations that are noble, altruistic, and of a spiritual nature can be so absorbed by the Higher Ego into eternity.

Human traits range from utter criminality to common humanity to supernal genius. How does Antahkarana feature in this? For the average person, Antahkarana disintegrates at death and its remains survive for a while as the Kama-rupa or, later, as the 'shell'. Ultimately, this bridge has to be destroyed or, rather, transcended during life by the merging of the personal nature with the Higher Self (see Figure 1 earlier).

In modern times it is the writer's persuasion that the monstrous crimes perpetrated against humanity by totalitarian dictators in full awareness of their actions but without any subsequent trace of penitence, may be due to the premature destruction of the Antahkarana. These are what is known in occultism as 'soul-less beings on earth'.

Some of the worst criminals in history have been men of great persuasion and high intellect—in the service of animal cunning. Such persons can be possessed of much charm and influence. It is the writer's contention that the extent to which a person displays cunning and manipulative characteristics is a measure of the extent to which such a person is an agent of dark forces. Such persons can be found in all walks of life, including spiritual societies.

Strengthening the Antahkarana is of paramount importance in awakening genius. Each step across the bridge is won by personal effort in disciplining and purifying the lower nature, which concomitantly increases the power and influence of the higher nature.

It is high time that criminologists, psychologists, psychiatrists, neurophysiologists, and religious ministers come to the stark realisation that occult science *alone* can explain such extremes of aberrant behaviour or genius that these mainstream disciplines, by their own admission, singularly fail to comprehend.

After-Death Communications

An ambitious two-year multilingual project, *Investigation of the phenomenology and impact of perceived spontaneous and direct After-Death Communications* (*ADCs*) was conducted by researcher, and expert on death-related experiences, Evelyn Elsaesser (*b.*1954) in order to widen the understanding and impact of such spontaneous communications. It identified seven principal types of communication:[19]

1. Sensing the familiar presence of a deceased family member or friend, without seeing or hearing them, or feeling a physical contact, or smelling a fragrance characteristic of the deceased.
2. Hearing a voice—either from an outside source, in the same way that a living person might be heard; or perceiving the communication without an external sound.
3. Feeling a physical contact—on a part of their body, for example, a touch, a pressure, a caress, a kiss, a hand placed on the shoulder or a real embrace.

4. Sensing the deceased—as apparitions that can occur indoors, for example at night in the bedroom; or outside, even in a car, on an aeroplane, etc.
5. Smelling a fragrance characteristic of the deceased—as an olfactory contact during which fragrances associated with a deceased person are perceived, like perfume, after-shave lotion, soap, tobacco, a characteristic body scent, etc.
6. During sleep—when falling asleep or waking up; or just falling asleep or just waking up.
7. Crisis within a 24-hour window before and after death.

The Silent Door of Eternity

Death, then, is the ending of the past caught in time. Truly great men have ever seen death in its rightful context—as transition and release and not extinction of consciousness—a portal to eternity.

When viewed from the higher vantage point of occult science, far from being a morbid or fearful affair, death is actually a sublime process in the continuity of consciousness and life. Death most certainly signifies an ending, but what exactly is it that ends? Only the past, meaning the ending of the known.

Too many mainstream scientists these days have allowed science to get in the way and blind their sensibilities to the subtler facts of consciousness. This happens when the lower mind blocks the light from the Higher Manas, resulting in wisdom becoming suffocated by excessive intellect.

Is the Brain a Computer? Are Humans Robots?

Although the brain does display some of the mechanical functions and characteristics of a computer, this is no reason to declare, as the majority of mainstream neuroscientists do, that the brain is nothing other than a 'wet computer'. It is minds and brains that created and produced computers, not the other way around.

All the definitions of computers have certain terminology in common: words and phrases like 'programmable', 'according to instructions given to it', 'written in an appropriate programming language', 'that stores programs and information' and 'that processes data according to a set of instructions'. All these terms make it patently obvious that a human programmer is involved. A computer cannot program itself; neither can it store programs and information by itself, nor can it alter data unless input by a human programmer. Computers (and machines in general) do only what humans program them to do.

The 'master analogy' unquestioningly accepted by the vast majority of mainstream scientists is: minds are to brains as software is to computers and minds cannot exist without brains just as software cannot exist without hardware.

Some computer scientists like the American, inventor and futurist Ray Kurzweil (*b.*1948) take this notion to extremes. He predicts that by 2029 computers will outsmart us and that the behaviour of computers will be indistinguishable from that of humans. And after the year 2045, Kurzweil maintains, machine intelligence will dominate human intelligence to such an extent that humans will no longer be capable of understanding machines. By then humans will have begun a process of 'machinisation', or 'transhumanism', which is the merging of humans and machines by cramming both human bodies and their brains with semiconductor chip implants, along with the fine tuning of their genetic material.[20]

It is reassuring to learn that a galaxy of scientists and entrepreneurs on the world stage have taken the step of pointing out their grave concerns about the threat posed by artificial intelligence (AI) and the ethical dilemma of bestowing moral responsibilities on robots. According to the American entrepreneur Elon Musk (*b.*1971) and others AI poses a greater threat to humanity than nuclear war.[21]

On the basis of such warnings, can all this hype from the aficionados of artificial intelligence and computationalism then be dismissed as the fantasies of nerds? We cannot do so because the latter have gained such prominence. Their ideas are highly pertinent to the whole question of the nature of consciousness and mind. We need to uncover and recover our humanness at all costs and the impending war against man- equals-computer, and then, computer-surpasses-man has to be fought in earnest.

The Computationalist's Strategy

The overriding aim of the computationalist is to eliminate all subjectivity because feelings are incompatible with the machine paradigm of man-equals-computer. And once the mind is reduced to a computer, all sense of personal responsibility, our pangs of conscience, our feelings for divinity and higher aspiration—all to do with being truly human—are eradicated or explained away at one stroke. The adopted strategy for doing so is in three stages.

Stage 1: Argue the case that man is just a computer and nothing more by ignoring everything that distinguishes man from a computer.

Stage 2: Eliminate feelings and subjective states, in which case man is no different from a computer. Can a computer feel anything?

Stage 3: If it can't be measured, it obviously doesn't exist, because science only recognises: accuracy which can be repeatably quantified over so-called truth which is a matter of subjective opinion; precision over meaning; quantity over quality; theory over experience; and establishment respectability over validity and evidence.

This strategy is implemented by way of a three-pronged attack against subjectivity using the interrelated arguments of:

1. roboticism or zombieism;
2. functionalism; and
3. brain states.

The world-renowned biologist Richard Dawkins (*b.*1941) has described human beings as 'lumbering robots'—which necessarily must also include himself in the epithet. He asserts: 'We are survival machines—robot vehicles blindly programed to preserve the selfish molecules known as genes.'[22] Such a 'truth' may fill us with sheer incredulity, even more so because he goes on to affirm that 'we animals are the most complicated things in the known universe.'[23] But we may not take issue with it. Nor are we allowed to ask who the blind programmer happens to be or what 'software' he used. Why?

Turning to functionalism, the important point is that subjectivity has been reduced to the physical, therefore to the observable and measurable. Once subjective states are dispensed with, one can indeed ask: 'What, then, is the difference between a man's brain and a computer? Can computers have feelings?'

Nor are we permitted to question the contention that there is no subjective state and that changes in brain states (i.e., physical mechanisms and processes in the brain) are the sole determining factor affecting our emotions and feelings.

In his book *The Master and His Emissary*, subtitled *The Divided Brain and the Making of the Western World*[24] the British academic Iain McGilchrist (*b.*1953) Fellow of All Souls College, Oxford, literary scholar, and Consultant Psychiatrist argues that the division of the brain into two hemispheres, whilst being essential to human existence, gives possibly incompatible versions of the world with quite different priorities and values. There are significant differences between the structure and function of the two hemispheres. However, as McGilchrist fully appreciates, to apportion specific brain functions exclusively to one hemisphere or another is an erroneous concept, as we now know that every type of function—including reason, emotion, language, and imagery—is subserved not by one hemisphere alone but by both.

This division of the brain into two hemispheres, leading to the divided nature of thought, corresponds on the physical plane to the occult teaching on the 'horizontally divided' double nature of the Mind Principle (the Higher Mind and the lower mind 'below' it). Crucially important is that *the hemispheres are more than mere machines with functions like computers. Instead, they are shown as underwriting whole, self-consistent versions of the world.*

McGilchrist's suggestion is that the encouragement of precise, categorical thinking at the expense of background vision and experience has now reached a point where it is seriously distorting both our lives and our thoughts. Our whole idea of what counts as scientific or professional has shifted towards literalism and precision—towards elevating quantity over quality, theory over experience, precision over truth—in a way that would have astonished even the seventeenth century founders of modern science, although they were already far advanced on that path. And the ideal of objectivity has developed in a way that would have surprised those early founders still more.

In his essay, *The Divided Brain and the Search for Meaning,*[25] McGilchrist asks why—despite the vast increase in material well-being—people are less happy today than they were around half a century ago. He suggests that the division between the two hemispheres of the brain has a critical effect on how we see and understand the world around us. The current left-brain-driven frenzy to convince us that we, thinking-feeling human beings, are just computing machines and nothing more is fitting proof of McGilchrist's prediction.

It is a tragedy, both for science and for humanity, that the science of polymaths, such as Leonardo da Vinci (1452–1519), has not managed to influence the ideas of later generations of thinkers or to stem the tide of the mechanistic science that emerged some two hundred years later. In the tradition of the great sages, da Vinci worked on the principle of the Hermetic philosophy, so whenever exploring the forms of nature in the macrocosm he looked for similar principles, patterns and processes in the human body—the microcosm. But then came the radically different mechanistic science of the *followers* of Descartes, Galileo, and also Newton.

A case in point is Newton, to whom the idea of the clockwork universe is attributed. But this just shows the ignorance and prejudice of those scientists and scholars who have, until recently, chosen to ignore Newton's colossal writings on alchemy and theology. With unmistakable clarity and force Newton proclaimed that nature is a living being and his writings glow with a love and reverence for deity, nature, and man considered as an organic unity.

This idea of matter being dead or inert and the body a machine, has persisted even to this day, especially in the context of the ever-increasing supremacy of scientific technology, whereas the spirit and soul aspects have either been conveniently ignored or explained away in terms of matter (such as by Francis Crick mentioned earlier). So, the root cause of the current malaise in thinking, epitomised by the computationalist cult, is this notion which now involves seeing everything natural as an object—inert, senseless and detached from us.

And such thinking might well explain the attitude that has led to the present ecological crisis whereby our planet is regarded as an inert object to plunder at our will for technological advancement and commercial expansion. By contrast, backed by sophisticated experimental evidence, quantum physics has repeatedly shown the interconnectedness of things in the world, the non-materiality of so-called physical matter and the role of consciousness in any consideration of quantum behaviour.

Yet the collective psyche of the vast majority of mainstream scientists, especially biologists, has hardly progressed beyond the outworn paradigm of the 'billiard ball' notion of dead matter behaving according to mechanical laws. Consequentially, life is reduced to the mechanical laws of classical physics without any higher informing principle. Indeed, it is the avowed aim of mainstream biology to reduce all things to physics and chemistry.

Mind and Cosmos by Thomas Nagel (*b.*1937), distinguished professor of philosophy at New York University,[26] is subtitled *Why the Materialist Neo-Darwinian Conception of Nature is Almost Certainly False*. This was guaranteed to produce howls of protest, which is precisely what happened. As his subtitle suggests, Darwinian evolution is insufficient to explain the emergence of consciousness. But even to question Darwinism politely is seen by the *cognoscenti* of science as an attack upon the omniscient 'God' of evolutionary theory. The resulting excommunication by the 'church of scientism' was swift and severe. Scientists are not physically burnt at the stake, but the personal abuse and the backlash against their careers is no different in principle.

Another fine example of the inviolability of the materialistic paradigm is the editorial that appeared on the front page of *Nature* under the title 'A book for burning?' The book in question was the first edition of *A New Science of Life*[27] by Rupert Sheldrake, which proposed the hypothesis of morphic resonance to explain the characteristic form and organisation of nature. The editor Sir John Maddox denounced it in a savage attack saying that 'even bad books should not be burned; works such as [Hitler's] *Mein Kampf* have become historical documents [...]. His [i.e., Sheldrake's] book is the best candidate for burning there has been for many years.'[28] Moreover, in a later BBC interview, Maddox said, 'Sheldrake is putting forward magic instead of science, and that can be condemned in exactly the language that the Pope used to condemn Galileo, and for the same reason. It is heresy.'[29]

From an esoteric perspective, these emotional counterattacks by the orthodox fraternity can be viewed as part of the backlash which is said to occur when one aspect of a universal principle is withdrawing and another is coming forward. When a new order is emerging, those from the earlier order cling on to habitual patterns to maintain status quo. Unable to move forward, any new insights or discoveries are forcibly quashed rather than carefully cherished. The fact that things have reached such a 'pitch' may indeed be regarded as an optimistic indicator of an incipient new paradigm for a new epoch. The current period of transition may well be seen by history as a crisis time.

Similar calumny and insult, but far more virulent, was heaped upon Blavatsky who also showed, with painstaking detail, that materialistic theories and Darwinian evolution were not wrong, as such, but wholly incomplete and therefore inadequate to account for consciousness, life, and evolution.

A Way Forward?

Longing to find beauty in what was, for him, an ugly and terrible world during the first Industrial Revolution, the legendary English Romantic poet John Keats (1795–1821) talks about what he calls 'negative capability'—the capacity to sit with the unknown, but with an inner conviction that something truly precious will come out of the unknown. In other words, to trust the process and have faith that there is a greater Consciousness that we can tune into.[30] But living with uncertainties and being with the unknown is something that the mainstream scientific and medical community find very intimidating and irritating. It underscores the deterministic characteristic of left-brain orientation towards precision, certainty, and objectivisation, whilst eschewing uncertainty, the unknown, and the unknowable.

But the right brain hemisphere, preferring to deal with wholes, rather than parts and with qualities over quantities, has no problem with subjective experience or waiting patiently in silence attuning to a higher power. So, this is the kind of holistic approach used by luminaries of science such as Leonardo and Newton—an approach that would help us greatly in facing the complex problems we face today.

Perhaps the deepest insights into the inner human condition and state (as opposed to the human body) come from sublime art, literature, poetry and music. Luminaries like Shakespeare, Keats, Mozart, Beethoven (1770–1827), Schubert (1797–1828), and Liszt (1811–1886) knew infinitely more about our minds, and what it means to be truly human, than establishment scientists and neurobiologists can ever hope to do.

Would the zealous protagonists of artificial intelligence who like to tell the world that we humans will be outstripped by androids in a matter of a few decades find the courage of their own convictions by being content with an android mate for an intimate and loving physical relationship, as an alternative, or in preference to a human being of flesh and blood?

It is incumbent upon us humans not to become robot-*like* as a result of rampant technology mechanising and de-humanising our lives. The real danger is not that robots will become more human-like but that *humans will become increasingly robot-like* as a result of over-reliance on technology, excessive artificial intelligence, mechanisation and the machine characteristic of existence. These are the tasteless fruits of a predominantly materialistic philosophy slowly eating into the spiritual core of our human life.

It is easy, then, to become intimidated by such scary stuff, especially so nowadays given the burgeoning transhumanist movement in science. However, the resolution rests in a simple appeal to common sense—common amongst so-called ordinary folk but a rare quality amongst the boffins of super-science because of the complete subjugation of wisdom by excessive intellect.

Man – a Mirror of Cosmos

Man, spiritual in his innermost self, expresses himself through the intermediate vesture of his soul and externally through a physical body. So, it is a grave mistake to conclude, as do the majority of scientists, that man is nothing more than a higher form of animal or, on the other hand, as gullible mystics do, that man is just a spiritual being, thus ignoring the basis of his earthly existence.

As a microcosm of the universe, man mirrors in his compound nature all elements, forces, and powers of the universe, the macrocosm. 'Man is essentially a permanent and immortal principle.' He is therefore not a compound, or aggregate, of two, three, five or seven entities but rather these represent the various *modes of consciousness* in which man can operate.

Man's composition has seven aspects and hence may be studied from seven different points of view. Accordingly, the clearest way to think of man is to regard him as one – Atma, the Divine Self, the true Self. Thus, man belongs to the highest region of the universe. Spirit is clothed in garment after garment, each garment belonging to a definite region of the universe and 'woven' from the substance thereof, enabling the Self to come into contact with that region, gain knowledge of it and work in it.

Man's consciousness can be said to be refracted into spiritual, mental, and material aspects, corresponding to the divine, psychic, and physical planes of nature.

In his book *Science and the Akashic Field,* Ervin László (*b.*1932), the distinguished Hungarian philosopher, gives robust evidence for consciousness beyond the brain, reincarnation and immortality in line with the latest theories about an *in-formation* or Akashic field ('A-field') whose intrinsic characteristic is holographic. Referring of course to science, he admits that: 'There is much that we do not yet understand about the farthermost reaches of human consciousness, but one thing stands out: consciousness does not vanish when the functions of the brain and body cease. It persists, can be recalled and, for a time at least, can also be communicated with. [...] The perennial intuition of an immortal soul is no longer inconsistent with what we are now *beginning to comprehend* through science about the true nature of reality [emphasis added].'[31]

Until science fully appreciates, and takes on board, the deep import of this core teaching of esotericism, its attempts to discover the stratospheres of consciousness—Universal Mind, the 'habitat' of the Higher Self — will be doomed to a squirrel-like progress around a wheel of cognitive science and brain research by whatever manner and means, like experimenting on hapless animals or with psychedelic drugs, all getting nowhere.

Occult science demands a *holistic* approach based on universality of enquiry, which clearly cannot be constrained to a single exposition, however erudite that may be. Hence, a one-track purist approach is neither sensible nor true to the spirit of the esoteric and occult tradition. Whereas the Theosophical teaching has constituted the backbone of our exposition, we have clearly stressed the importance of teachings from the worldwide perennial wisdom tradition.

To maintain a doctrinaire attitude towards any one system of classifying man is wholly obtuse. Let us constantly remember that we are dealing with forces and states of consciousness and not with water-tight compartments. Sadly, within the Theosophical Society much time and energy has been wasted on tedious squabbling over which occult system is the authentic one. Of course, as an organisation, it is by no means unique in this respect. Such feuds are born of a literal-minded, bookish attitude instead of understanding the context and reasons for the different ways of classifying man's principles and subtle bodies. This attitude has no place for the aspirant whose only concern should be with truth wherever it may lead and from whatever source.

NOTES

1. 'Mahatma Gandhi Quotes' <https://www.azquotes.com/author/5308-Mahatma_Gandhi> accessed 14 May 2020.
2. BBC News, 'UK Most Overweight Country in Western Europe says OECD [Organisation for Economic Co-operation and Development]', 11 November 2017 <https://www.bbc.co.uk/news/ uk-41953530> accessed 14 May 2020.
3. 'Dark Energy, Dark Matter', NASA Science Newsletter <https://science.nasa.gov/astrophysics/focus-areas/what-is-dark-energy> accessed 8 February 2020.
4. *KT*, 'The Septenary Nature of Man', 91.
5. G. de Purucker, *Occult Glossary* (Pasadena, California: Theosophical University Press, 1996), 131.
6. Reworded from *KT*, 'On the Various Post-Mortem States', 121–2.
7. Marginally reworded from *CW*-XII, 'Instruction No. III', 608.
8. *TSGLOSS*, 74.
9. H. P. Blavatsky, *Studies in Occultism: Practical occultism* – a Collection of Articles from *Lucifer*, H. P. Blavatsky's magazine, between 1887–1891 (Montana, US: Literary Licensing, 2014).
10. Ernest Shackleton, *South: The endurance expedition* (London: Penguin Classics, 1914), 204.
11. Patricia Pearson, *Opening Heaven's Door: What the dying are trying to say about where they're going* (UK: Simon & Schuster, 2014).
12. See *STA*, 'The Mysteries and Their Emissaries', CXCIX.
13. David Lorimer (ed.), *Prophet for Our Times: The life and teachings of Peter Deunov*, foreword by Wayne W. Dyer (London: Hay House, 2015), 24. See also pp. 19–20, 25–6.
14. Peter Deunov, 'Divine Providence', translations from Bulgarian of Authentic Versions of the Lectures and Other

Works (19 August 2008) <http://www.beinsa-douno.net/lectures/Divine- Providence.html> accessed 8 February 2020.

15. Geoffrey Barborka, *The Divine Plan* (2nd edn, rev. and enl., Adyar, Madras: Theosophical Publishing House, 1964; repr. 1980), 406 n.
16. See Adam Warcup, 'An Inquiry into the Nature of Mind', the Blavatsky Lecture, 1981 (London: The Theosophical Society), 18.
17. An entirely safe and meaningful account in terms of the oscillation of consciousness between two focal extremes is in James Perkins, *A Geometry of Space-Consciousness* (1964; rev. and enl. 3rd edn, Adyar, Chennai: Theosophical Publishing House, 1977, repr. 2004).
18. *Theosophical Encyclopedia,* ed. Philip S. Harris, Vincente R. Hao Chin, Jr., and Richard W. Brooks (Philippines: Theosophical Publishing House, 2006), 38.
19. Evelyn Elsaesser, *Spontaneous Contacts with the Deceased* (UK: John Hunt Publishing, 2023).
20. David Gelernter, 'The Closing of the Scientific Mind'.
21. *The Guardian*, 27 October, 2014.
22. Richard Dawkins, *The Selfish Gene* (Oxford: Oxford University Press, 2006), xx.
23. ——*The Blind Watchmaker* (New York: W. W. Norton & Company, 1986), 1.
24. Iain McGilchrist, *The Master and His Emissary: The divided brain and the making of the Western world* (2010; rev. and enl. 2nd edn, New Haven and London: Yale University Press, 2019).
25. ——*The Divided Brain and the Search for Meaning* (New Haven and London: Yale University Press, 2012).
26. Thomas Nagel, *Mind and Cosmos: Why the materialist neo-darwinian conception of nature is almost certainly false* (Oxford: Oxford University Press), 2012.

27. Rupert Sheldrake, *A New Science of Life: The hypothesis of formative causation*, Blond and Briggs, 1981.
28. J. Maddox, 'A Book for Burning?' *Nature*, 293 (1981), 245–6.
29. Tony Edwards, Heretic: *Rupert Sheldrake*, BBC 2 (19 July 1994) <https://genome.ch.bbc.co.uk/f2ff3746e43e472fb35c4a7ceb68e2e8> accessed 19 May 2020.
30. John Keats, 'Selection from Keats's Letters (1817)', *Poetry Foundation* https://www.poetryfoundation.org/resources/learning/essays/detail/69384 accessed 8 February 2020.
31. Ervin László, *Science and the Akashic Field: An integral theory of everything* (2004; 2nd edn, Rochester, Vermont: Inner Traditions, 2007), 128.

Part Four

Instruments of Perception

Everything in the Universe, throughout all its kingdoms, is conscious: i.e., endowed with a consciousness of its own kind and on its own plane of perception. We men must remember that because we do not perceive any signs—which we can recognise—of consciousness, say, in stones, we have no right to say that no consciousness exists there. There is no such thing as either 'dead' or 'blind' matter, as there is no 'Blind' or 'Unconscious' Law. Occult philosophy never stops at surface appearances, and for it the noumenal essences have more reality than their objective counterparts.

H. P. Blavatsky, *The Secret Doctrine*[1]

What is the one central factor that clouds a person's awareness of the deeper realms of existence? It is not difficult to identify. Materialism is a contributory factor, but the central one is, quite simply, worldly weariness. Excessive involvement and engagement with the material world, over and beyond the natural requirements of physical life, induces that inner fatigue of mind that dulls sensitivity and clouds perception. Fuelled by the internet, we live in an age where it becomes increasingly difficult to discern wisdom from the tsunami of knowledge produced by avalanches of data.

As far back as the eighteenth century, the English Romantic poet William Wordsworth (1770–1850) foresaw the ongoing pattern whereby excessive technology was slowly, but inexorably, eating into the human spirit and 'roboticising' the human being of flesh and blood. The world is nowadays awash with data, information, and knowledge. This is the conventional outlook on education in schools and universities. Yet what is needed, more urgently than ever, are transformational learning processes that empower people to engage in personal transformation. This will lead to wisdom that provides the rudder to navigate skilfully through the current changeable and unsettling times.

There is something which lies beyond the duality and ephemeral nature of everyday life. Though the experience of it may be feeble for the majority of people, yet the dim remembrance still persists on the periphery of awareness, whereupon illusory attachment weakens. During those instants it makes its presence known in a brightened perception of the present moment.

A 'Knew' Way of Knowing – a New Way of Looking

Although the voices of religion, philosophy, and the best of modern science in the outer world are calling out for us to deepen our awareness of the Invisible Higher Power, the only authentic call which we can truly hear and respond to is the inner voice of our own Higher Self. In its elusive tones, it continually whispers to us through the intervening layers of our business and worldly involvement to eschew transitory pursuits and look beyond illusion.

Symbolism – the Language of the Mystery Teachings

Why does the literature of religion, esoteric philosophy, and occultism abound in the use of figurative language? Ordinary language wedded to a linear, intellectual mode of thought is well suited to conveying precise information, objective facts and scientific theories. But it is a poor medium for conveying the subjective realms of experience and deep truths that are, necessarily, beyond words and above logical reason. Recourse to parables and allegories, metaphors and similes, and, above all, symbols and analogies is the only means of unlocking the meaning behind what is in truth, ineffable.

A *proverb* is a short, pithy saying in common use, held to embody a general truth. A *parable* is a short narrative of imagined events intended to illustrate a moral or spiritual lesson. It differs from a *fable* which employs animals, plants, inanimate objects, or forces of nature as characters, whereas parables have human characters. Parables, then, are the general means of conveying the esoteric meaning in exoteric parlance—therefore, to be understood and told to the masses. However, that does not mean that so-called common folk do not have 'eyes to see' and 'ears to hear' and therefore that parables cannot be understood at different levels.

An *allegory* is a vehicle for conveying truth by representing an abstract or spiritual meaning through concrete or material forms. The subject under question is treated in the guise of the narrative of another, such as in a play, story or poem. For example, the Cretan legend of the Minotaur is an allegorical depiction of the unbridled, carnal passions in man under the guise of a mythological creature with the head of a bull and the body of a man.

Regarding *similes*, the emphasis is to aid the understanding of a subject by using the words 'like' or 'as' to compare the subject to something else. A *metaphor* is similar to a simile but compares the subject to something else without an illustrative case: the comparison is implied or stated directly. In both cases a descriptive term is applied in the imaginative but not literal sense.

We now come to, arguably, the most significant forms of figurative languages used in esotericism and occultism—*analogies* and *symbols*. In an analogy there is a correspondence, or partial similarity, between two or more things. *They are related by a common principle*.

A symbol is something (such as a figure or character or picture) regarded as typifying something else, especially in the nature of an idea or quality. Musical notation and the letters standing for the chemical elements are obvious examples. Symbolism is the pictorial representation of an idea or a thought. Archaic writing initially had no characters, but a symbol generally stood for a whole phrase or sentence.

For the student of the Mystery Teachings symbols are a most potent means of unlocking hidden depths of meaning and releasing intuition. The reason for this is that the meaning attaching to a symbol may be taken and understood on many levels—physical, intellectual, spiritual—and what is revealed is entirely dependent on the context and, crucially, the degree of awakened intuition of the seeker.

One such example of a powerful symbol is the *ouroboros* depicting a snake swallowing or biting its own tail. The basic idea is that of constant regeneration and eternal cyclicity of life through the renewal of forms.

While the greatest minds of the Jewish and Christian worlds have realised that the Bible is a book of allegories, few seem to have taken the trouble to investigate its symbols and parables. When Moses instituted his Mysteries, he gave to a chosen few initiates certain oral teachings which could never be written down and could only be preserved from one generation to the next by word-of-mouth transmission. Those instructions were in the form of philosophical keys, by means of which the allegories were made to reveal their hidden significance. These mystic keys to their sacred writings were called by the Jews the Kaballah (or Qabbalah or Cabala).

The modern world seems to have forgotten the existence of those unwritten teachings which explained satisfactorily the apparent contradictions of the written Scriptures.

Symbolism is the most secret and the most enduring of all mysteries. Symbols have an in-built elasticity of meaning in many dimensions. In addition to visual symbolism, there is also a profound symbolism hidden in theatre and music drama. The various characters in Shakespeare's plays or in the dramatic operas of Mozart, Beethoven, and Wagner (1813–1883), symbolise archetypal forces in nature and aspects of the human psyche and epitomise the drama and struggle of the human spirit in conquering adversity in its journey towards enlightenment.

A few examples are now provided of powerful symbols that have withstood the test of time and have assumed a virtually archetypal significance in the collective psyche of mankind.

The *thread* is a common symbol of the philosophical quest for truth through the labyrinth, traps and woes of mortal existence. In the famous Cretan legend, the hero Theseus unrolls a ball of thread as he winds his way into the labyrinth to encounter and slay the Minotaur and then follows the thread back out of the maze. The other end of the thread is held firm by Princess Ariadne who falls in love with the handsome youth.

A different allegorical significance is the *sacred thread* that binds together the perennial wisdom, connecting disparate disciplines, such as science, religion, and philosophy, like beads on a necklace and forming them into an organic unity.

The *labyrinth* is a symbol from the West for depicting the ordinary, mortal existence. The popular meaning, intuitively suggested by both symbols, is the confusion and perplexity of everyday life invariably devoid of illumination to reveal the straight and narrow path ahead.

When we enter into terrestrial life we are indeed in a labyrinthine maze of all kinds of experiences. We do not know which of many paths to take and we have to learn to control our personal Minotaur so that its sense-driven appetites do not get the better of our higher nature. When we succeed in disciplining our lower nature, our path ahead becomes clarified. The labyrinth is thus the symbolic pattern of our coming into and out of birth.

The *forest,* a common symbol from the East, has the obvious connotation of mental confusion and so not 'seeing the wood for the trees'. But there is the added meaning of danger from poisonous serpents, deadly beasts, and wild demons which all symbolise the untamed passions and ungoverned thoughts which literally, like demons, assail our equanimity and obstruct our life mission. We find the forest mentioned in several Indian legends.

The symbol of the *fish* is found in the sacred traditions of both East and West. In India, *Matsya* is the avatar of Vishnu who takes the form of a fish and forewarns the primordial human race about an impending flood and orders him to collect all grains and living creatures to be preserved in a boat. The remarkable similarity of this legend to that of Noah's Ark is another fitting demonstration of the self-consistency and universality of symbolism used to elucidate esoteric knowledge and spiritual truths.

In the Druidic traditions the *salmon* is a symbol of constant renewal and seeking the source of all existence.[2] Salmon return to rivers from the ocean and swim upstream to their original hatching place to lay and fertilise their eggs before they die.

Symbols of the Tarot

The Tarot cards represent the elements of life, philosophy and consciousness and its expression. We will explore the symbolism in three of the trump cards from the Major Arcana as depicted in Figure 2 below.

Image Credit: Pamela Coleman Smith/https://en.wikipedia.org/wiki/Rider-Waite_tarot_deck

Figure 2: Trump Cards from the Major Arcana

The Hermit – Concealing the Light of Wisdom

The ninth numbered trump card is called *The Hermit* and portrays an aged man, robed in a monkish habit and cowl, leaning on a staff. He personifies the secret organisations which for countless centuries have carefully concealed the light of the Ancient Wisdom from the profane. The hermit's staff denotes the mind principle and is therefore knowledge.

He shields the lamp—representing the Divine Self—behind a rectangular cape—man's physical body—to emphasise that wisdom once exposed to the fury of ignorance is extinguished like the tiny flame of a lamp unprotected from the storm. Man's bodies form a cloak through which his divine nature is faintly visible like the flame of the partly covered lantern. Through renunciation—the Hermetic life—man attains depth of character and tranquility of spirit.[3]

The Lovers – the Price of Free Will

The sixth trump card is *The Lovers,* portrayed by a youth with a female figure on either side. The maidens represent the twofold soul of man (spiritual and animal), the first his guardian angel, the second his ever-present demon (or irrational soul), *and both love him*. The youth stands at the beginning of mature life where he must choose between virtue and vice, the eternal and the temporal. Above, in a halo of light, is the genius of fate (his star), mistaken for Cupid by the uninformed. If the youth chooses unwisely, the arrow of blindfolded fate will transfix him.

This card reminds man that the price of free will—or, more correctly, the power of choice—is responsibility.[4]

It illustrates a cardinal teaching of occultism, namely, that the dual nature of the mind principle has the power of two choices: either to 'rise' and align with the guardian angel, whereupon all is well in accordance with divine law, or to 'fall' and attach to the demon.

The Judgement – Liberating the Spiritual Nature from the Material Sepulchre

The twentieth major card is *The Judgement* and portrays three figures rising, apparently, from their tombs though only one coffin is visible. Only one of the three figures is actually rising. This is because one-third of the spirit actually enters the body whilst the other two-thirds, constituting the overman, remain in the heavens. Above them is a winged figure (presumably the Angel Gabriel) blowing a trumpet.

This card represents the liberation of man's threefold spiritual nature—Atma-Buddhi- Manas—from the sepulchre of his material constitution upon death for the average person, but in life for the illumined sage. The blast of the trumpet represents the creative word which liberates the soul from her terrestrial limitations.[5]

The Game of Chess – the Drama of Life

In its symbolism *chess* is the most significant of all games. It has been called 'the royal game'—the pastime of kings. Like the Tarot cards, the chessmen represent the elements of life and philosophy, consciousness and its expression. The game was played in India and China long before its introduction into Europe. East Indian princes were wont to sit on the balconies of their palaces and play chess with living men standing upon a checker-board pavement of black and white marble in the courtyard below.

The chessboard consists of sixty-four symbolising the floor of the House of the Mysteries where strangely carved figures moved according to fixed law and conducted the ongoing great war between light and darkness. The king represents the spirit, the queen the intuitive mind, the bishops the emotions, the knights the vitality, the castles or rooks, the model and physical body. The pieces upon the king's side are positive. Those upon the queen's side are negative. The pawns are the perceptive faculties and sensory impulses—being aspects of the soul. The white king and his suite symbolise the self and its vehicles; the black king and his retinue, the false ego. The game of chess concerns the eternal struggle of each part of man's compound nature against the shadow of itself.

It is not so far-fetched to mention that the more the governance of nations is organised along the above lines of hierarchy, the more stable and enduring will be its society.

Alchemical Secrets and the Philosopher's Stone

This is not the place to add to the colossal tracts that have been written for many centuries about alchemy in the West and the meaning of the Philosopher's Stone, other than its symbolic importance. Under the symbolism of an alchemical marriage, the Rosicrucians and mediæval philosophers concealed the secret system of spiritual culture whereby they hoped to co-ordinate the scattered fragments of both the human and social organisms. The bigotry of the church, the tyranny of the state and the fury of the mob are the three murderous agencies of society which seek to destroy truth.

The true Rosicrucian Brotherhood consisted of a limited number of highly developed initiates or adepts, those of the higher degrees being no longer subject to the laws of mortality and candidates being accepted into the Order only after long periods of probation. Adepts possessed the secret of the Philosopher's Stone and knew the process of transmuting base metals into gold. But they taught that these were only allegorical terms concealing the true mystery of human regeneration—the transmutation of the 'base elements' of man's lower nature into the 'gold' of intellectual and spiritual realisation.

Those who have sought to penetrate the secrets of the Philosopher's Stone have invariably failed because they approached their subject from a purely materialistic angle. So, for instance, if an alchemist gave directions for the special treatment of Sulphur, Mercury, and Salt, with the assertion that by carrying out these directions properly, one would obtain Aurum (gold), *he really spoke of a method to direct the thinking, feeling, and willing activities of the soul in such a way as to gain true Wisdom.*[6]

The Misunderstanding and Misuse of Allegories and Symbols

There is, arguably, no finer example of the problems (ranging from acrimony between friends to outright war) caused by a dead-letter, literal interpretation of scriptures than the fate suffered by the Holy Bible. The greatest minds of the Jewish and Christian worlds have realised that the Bible is entirely a book of allegories. It has always been the case that the highest teachings were never written down but preserved from one generation to the next by word-of-mouth transmission.

These instructions were in the form of philosophic keys, by means of which the allegories were made to reveal their hidden significance as when Moses instituted his Mysteries to elected initiates through oral teachings which could never be written down. These mystic keys to their sacred writings were called by the Jews, the Kabbalah. Then, in the East, the sacred Vedas were taught orally and only written down in the form of the Upanishads, Brahmanas, and Puranas by later generations. Also, the early literature of Buddhism was orally transmitted.[7]

Animal and Human Sacrifice

A common jibe hurled at the pagans and ancient cultures, such as the Incas, Aztecs, Hawaiians, and Egyptians, is that in order to propitiate and appease the gods they sacrificed animals and, in some cases, humans. This accusation is not without some justification, but it displays complete ignorance regarding the status of ancient cultures during their zenith and in their decline. It was only during the later period of degeneration and decay that the eternal truths were progressively corrupted and carnalised. Hence, unable to understand its metaphysical or philosophical import, the symbolic meaning of sacrifice was taken literally and materialised into a physical ritual perpetrated upon hapless animals or humans.

A candidate, when first entering the precincts of the sanctuary, must offer upon the altar not a poor unoffending bull or ram *but its correspondence within his own nature*. The bull, being symbolic of earthiness, represented his own gross constitution which must be burned up by the fire of his divinity. The sacrificing of beasts, and in some cases human beings, upon the altars of the pagans was the result of their subsequent ignorance concerning the fundamental principle underlying sacrifice. *They did not realise that their offerings must come from within their own natures in order to be genuine.*[8]

We are entitled to point out that those who sneer at such ancient cultures, because of their supposedly uncivilized ignorance, might wish to reflect on the barbaric brutalities perpetrated by modern, 'civilized' man—such as in Nazi Germany, Communist Russia and other dictatorial regimes as in Chile and Iraq. Human nature at its vilest is ever the same—in any era, ancient or modern.

Secrets of the Zodiac

Man has ever gazed heavenwards to contemplate and acknowledge his kinship with the stars. Even the most hard-nosed astronomers will admit that something inexplicably mysterious moves them to study and map the heavens above. Why, then, this fascination with the celestial spheres, which they so confidently declare to be just lifeless physical matter in a purposeless universe?

In astronomy, the ecliptic is the great circle traced out by the Sun on its annual path against the stellar background. It is essentially the plane of the Earth's orbit around the Sun.

The ecliptic forms the centre of an imaginary band of sky, approximately 18 degrees of arc wide, known as the zodiac, on which the Sun, Moon, and all the solar system planets are seen always to move. This band is divided into twelve equal parts, each a segment of sky 30 arc degrees wide, and each named after a prominent constellation situated in it.

These are the signs of the zodiac, the Sun appearing to move through these signs at the approximate rate of one per month. The signs give us a clue about some of the terminology used today. The first point of Aries was named when the vernal equinox was actually in the constellation Aries, when the Greek astronomer and mathematician Hipparchus of Nicaea (190–120 BC) defined it in 130 BC. It has since moved into Pisces.

The word 'zodiac' has a dual meaning. It may refer to the fixed or intellectual zodiac containing the fixed stars or else to the movable or natural zodiac containing the moveable or wandering stars.

But why is it that so many signs of the zodiac are named after animals—ram, bull, lion, etc.? Is it just because the outline of the constellations can with a little imagination be traced to the figures of animals? Or is there a deeper reason? Interestingly and significantly, the word 'zodiac' is derived from the Greek word *zodion*, a diminutive of *zoon* meaning animal and here we have our first clue about the subjective influences and correlations concerning the stars and planets. For animals are not dead things. *They typify energies, each according to its nature* (like the resolute endurance of the bull or the noble courage of the lion).

The zodiac was known in India and Egypt for incalculable ages and the knowledge of the sages of these countries with regard to the occult influence of the stars and heavenly bodies on our planet was far greater than profane astronomy ever admits.

Astrology – Royal Science or Pseudoscience?

Virtually without exception, the most prominent and internationally fêted scientists of today regard astrology as a 'pseudoscience', stated politely, or 'utter nonsense', put bluntly. They are not entirely wrong if they base their opinions just on the puerile columns of newspaper astrology without bothering to dig deeper as genuine scientists ought to do. But are these modern pundits missing something?

What is a pseudoscience? In short, it is what appears to be scientific but in fact is not: the mask of science rather than science itself. More technically, according to the Oxford English Dictionary (ninth edition, 1995), pseudoscience is 'a pretended or spurious science; a collection of beliefs mistakenly regarded as based on scientific method.'

There is a knowledge and knowing that can be derived by means not amenable to physical science. And that knowledge, central to the science of astrology, is transcendental metaphysics dealing with the greatest and most abstruse problems concerning the universe and man—a far cry from pure materialism. And materialistic science is singularly ill-equipped to deal with subjects that lie outside its self-imposed boundaries of investigating objective and physical phenomena.

Man Is the 'Measure of All Things'

Other than with individual practitioners, psychiatry as a whole has not achieved the same success rate as physical medicine has using drug therapy and surgery. Why? Because ailments of the mind involve a profound understanding of the human state and condition, not only the neurological processes in the brain. And to the extent that psychiatry divorces itself from spirituality, concentrating on outward behavioural characteristics and treating brain disorders through drugs, its attempts to treat the mind as a whole will be vitiated.

The zodiac was mentioned as one such universal symbol. We now focus on the human body, also one of the oldest, most profound, and universal symbols. According to Protagoras (490–420 BC), the pre-Socratic Greek philosopher: 'Man is the measure of all things.'[9]

The esoteric philosophers from antiquity to the present day recognised the futility of attempting to cope intellectually with that which transcends the comprehension of the rational faculties, and so invoked the well-known Hermetic Axiom, 'As above, so below'. This was an attempt to understand the inconceivable Divinity by studying Its reflection in man.

This is why the Greeks, Persians, Egyptians, Hindus and, even, the Gnostics (before the Orthodox Church suffocated the esoteric dimension of Christianity) considered a philosophical analysis of man's three-fold nature to be an indispensable part of ethical and religious training. The Mysteries of every nation taught that the laws, elements, and powers of the universe were epitomised in the human constitution and that everything which existed outside of man had its analogue within him.

From Exaltation to Idolisation

Long before idolatry corrupted religion, statues of men were placed in the sanctuary of the temple, the human figure symbolising the divine power in all its intricate manifestations. The hierophants accepted man as their textbook and through the study of him learned to understand the greater and more abstruse mysteries of the celestial scheme of which they were a part. After ages of investigation the manikin became a mass of intricate hieroglyphs and symbolic figures. Every part had its secret meaning. The measurements formed a basic standard by which it was possible to measure all parts of the cosmos. It was an organic and composite emblem of all the knowledge possessed by the sages and hierophants.

In vain did the initiates of old warn their disciples that an image is not a reality but merely the objectification of a subjective idea. The images of the gods were not designed to be objects of worship but were to be regarded merely as emblems, or reminders, of invisible powers and principles. Similarly, the body of man was not to be considered as the individual but only as the house of the individual. In a state of grossness and perversion, man's body is the tomb, or prison, of a divine inhabitant but in a state of refinement, the temple of his spirit.

Then came the age of idolatry. The Mysteries decayed from within. The secrets were lost and none knew the identity of the mysterious man who stood over the altar. It was remembered only that the figure was a sacred and glorious symbol of a great and universal Power and it finally came to be looked upon as a god—the one in whose image man was made. Having lost the knowledge of the purpose for which the manikin was originally constructed, the priests worshiped this effigy until finally their lack of spiritual understanding brought the temple down in ruins about their heads and the statue crumbled with the civilization that had forgotten its meaning. By progressive stages, the living esoteric meaning was starved in proportion to the increasing adulation of the outward, exoteric form.[10]

The Three Grand Centres of Consciousness-Power

The ancient wisdom teaches that all bodies—whether spiritual or material—have three centres called by the Greeks the *upper,* the *middle,* and the *lower* centres. While that which is above is generally considered superior in dignity and power, in reality that which is in the centre is superior and anterior to both that which is said to be above and that which is said to be below.

Since the superior (or spiritual) centre is in the midst of the other two, its analogue in the physical body is the heart—the most spiritual and mysterious organ in the human body. The second centre is elevated to the position of greatest physical dignity—

the brain. The third (or lower) centre is relegated to the position of least physical dignity but greatest physical importance—the generative system. Thus, the heart is symbolically the source of life; the brain, the link by which, through rational intelligence, life and form (spirit and matter in the generic sense) are united; and the generative system the source of that power by which physical organisms are produced. The ideals and aspirations of the individual depend largely upon which of these three centres of power predominates. In the materialist the lower centre is the strongest, in the intellectualist the higher centre, but in the initiate the middle centre controls wholesomely both the mind and the body.[11]

Here we may comprehend the meaning of those immortal words: '*As a man thinketh in his heart, so is he.*' As there are seven hearts in the brain so there are seven brains in the heart, but this is a matter of super physics of which little can be said at the present time.[12] Nonetheless, this writer maintains that there is no better policy for the modern scientist to follow than '*to think with the heart, and feel with the brain*'.

The Heart Rules the Head

In a notable paper on occult physiology, Blavatsky explains that there are seven cavities in the brain which, during life, are empty in the ordinary sense of the word, but in reality, are filled with Akasha, each cavity having its own colour according to the state of consciousness of the individual. These brain cavities are referred to in occultism as the 'Seven Harmonies' *and it is in these that visions must be reflected, if they are to remain in the brain-memory.*[13] But reflection surely implies a source of light? Where does that light come from? Blavatsky continues by pointing out that the seven cavities are the parts of the brain that receive impressions from the heart and enable the memory of the heart to be impressed on the memory of the brain.

In essence, the fountainhead of consciousness in man is the heart and not the brain whose memory (light) is borrowed from the heart.

What is the crux of the recent discoveries in science about the human heart? In the article *'Is God All in Your Head?'*, Craig Hamilton, a leader in the emerging field of evolutionary spirituality, comments: 'The cranium may be home to the smartest organ in town, but when it comes to sheer magnetism, the gray matter in your head may have a little competition on its hands. According to the new science of neurocardiology, we have a *second brain* in the form of a dense cluster of *neurons,* in the *heart,* and its electromagnetic field is five thousand times stronger than the brain upstairs. So, don't be surprised if the next person telling you to 'follow your heart' is your doctor [writer's emphases].'[14]

Current research data from the HeartMath Institute Research Centre (an internationally recognized global leader in emotional physiology, optimal function, resilience, and stress-management research) puts the value of the electromagnetic field strength of the heart to around one thousand times stronger than that generated by the brain.'[15]

It is important to distinguish between the 'second brain in the form of a dense cluster of neurons, in the heart,' from what is now commonly referred to as the 'second brain', or 'brain in the gut'. This is the enteric (relating to, or occurring in, the intestines) nervous system comprising two thin layers of more than one hundred million nerve cells lining the gastrointestinal tract from the oesophagus to the rectum.[16]

Occultists have always known that consciousness is ubiquitous and pervades each and every particle of the universe as indeed it does the human body and its cells. The medical profession uses the safe term 'cellular memory' to define the idea that the cells in our bodies contain information about our personalities, tastes and histories. Evidence of this phenomenon has been found to be prevalent in heart transplant recipients and cannot readily be explained away by the influence of the drugs used to suppress donor rejection.

A striking example concerns the heart of a murdered ten-year-old girl, which was transplanted into an eight-year-old girl who was subsequently able to provide such accurate details about the crime as to lead to the arrest and conviction of the murderer.[17] However, explanations other than cellular memory cannot be ruled out; for example, what is commonly referred to as 'spirit attachment' or the sudden onset of cases suggestive of reincarnation where there is no question of transference of physical cells between the bodies of different individuals widely separated in distance and time.

The HeartMath Institute Research Center conducts basic research into psychophysiology, neurocardiology and biophysics in collaboration with universities, research centres and health-care partners. One of the primary areas of focus is the mechanisms by which emotions influence cognitive processes, behaviour, health, and the global interconnectivity between people and the Earth's energetic systems. What is of great importance is how this research has significantly advanced the understanding of heart-brain interactions.

The human heart is evidently far more than just a biological pump. But any lingering doubts in the mind of the reader should be dispelled by recent and compelling evidence from professor of pathology, Asok K. Mukhopadhyay and cardiologist Jay Relan, cited in their paper *The DeepScience of NeuroCardiology*. 'Contrary to present mainstream science which has accepted [the] heart as a blood-pumping machine working under [the] nervous system, the popular belief system reigns from time antiquity that the heart is the seat of feelings, which participates while making decisions.'[18]

The authors argue that the current dichotomy between the brain and the heart could be the principal cause of needless human suffering. Moreover, the heartless pursuance of science, outmoded religious rituals, and politics are the causes of environmental, economic and human disasters. The remedy lies in a union of the heart and the brain of mankind leading to an integral evolution which embraces the cerebral aspects of the brain as well as the compassion and empathy of the heart.

A scientific approach to this issue is, however, essential and the disciplines of cardiology, neurology, neurocardiology, psychology, and sociology should be nurtured as, collectively, they represent the science of consciousness. Such a joint endeavour would actuate a meaningful co-existence of neuro-centric with cardio-centric consciousness, rooted in the full realisation of the primacy of consciousness.

The brain, heart, gut, proteins, DNA, etc. enable us to express human consciousness whilst on Earth. All these factors play their respective roles in the emergence from the present *Homo sapiens* of the '*Homo spiritualis*' with qualities such as love, trust, and hope centred in the heart. As the French philosopher Pierre Teilhard de Chardin (1881–1955) commented, 'We are not human beings having a spiritual experience. We are spiritual beings having a human experience.'[19]

Significantly, we are reminded of how Einstein took serious note of the inadequate development of the *affective* brain (responsible for attitudinal development of personality) in science and education: 'The release of atom power has changed everything except our way of thinking ... the solution to this problem lies in the heart of mankind. If only I had known, I should have become a watchmaker.' (1945)[20]

Symbolic Representations of the Occult Powers in Man

Throughout the ages, the latent powers in man have been symbolised in various ways such as mythological creatures, half human and half beast, or through tales of heroic struggles. The wooden horse of Troy is an excellent example of an exoteric narrative covering a hidden, esoteric meaning.

Ironically, the story itself is in the nature of a Trojan Horse because once the exoteric layers are peeled away, a myriad of meanings burst forth, both metaphorically and esoterically. There are four key themes. In sequential order they are: transgression, war, subterfuge, and victory. What do they imply?

Metaphorically, the Trojan Horse signifies any trick or subterfuge that causes an innocent victim to invite an enemy into its own protected environment. A computer virus which tricks users into willingly running a malicious software program is a modern example of the malevolent aspect of the Trojan Horse.

Esoterically however, the Trojan Horse symbolises the occult powers in man. The wooden horse represents the body of man containing those infinite potentialities which will later issue forth and conquer his environment. The siege of Troy is a symbolical account of the abduction of the spiritual soul—Helene—by the personality—Paris and its final deliverance through persevering struggle.

It conveys the message that final redemption can only occur through the secret doctrine (meaning through the teachings of occult science) but never through science and intellect alone unaided by soul wisdom. When wisdom and intuition are strangled by unbridled materialism, only a titanic effort will release the suffocating grip. And that will occur only, it bears repeating, through enlightened science wedded to its ally, the secret doctrine which alone can dispel that greatest disease of all—ignorance. But ignorance of what? Ignorance of our true and inner nature.

Awakening Occult Powers

The Mysteries taught the worthy candidate about the unfolding of his innate nature and the natural ripening of his occult powers as a logical outcome. He was shown how to awaken the forces lying dormant within certain physical organs, channels and centres in the body which are the veils of spiritual centres of forces and powers. What these were, and how they could be awakened to activity, was never revealed to the unregenerate, for the philosophers realised that once a man understood the complete working of any system, he could accomplish a prescribed end without being qualified to manipulate and control the effects which he has produced. Goethe's tale of 'The Sorcerer's Apprentice', is about an old sorcerer leaving behind his apprentice in the shop to undertake chores. The apprentice then decides to use magic powers to do the chores for him but mayhem results from evoking such unseen powers with only a fragment of know-how, thus lacking full knowledge about their control. For this reason, long periods of probation were imposed so that the knowledge of how to become 'as the gods' might remain the sole possession of the worthy.

Lest knowledge of the Mysteries be lost, it was concealed in allegories and myths, meaningless to the profane but self-evident to those acquainted with that theory of personal redemption which was the foundation of philosophical theology. The entire New Testament is, we are told, an ingeniously concealed exposition of the secret processes of human regeneration. The characters, both pleasant and unpleasant, so long considered as historical men and women are really the personification of certain processes which take place in the human body. This happens when man begins the task of consciously liberating himself from the illusion of materialism, its associated bondage to ignorance and death can be regarded as the extinguishing of consciousness. In modern times, in step with the enormous discoveries of science, a portion of this knowledge was promulgated (without personifications) through the modern Theosophical movement.

Symbols of Man's Principles

In the Esoteric Philosophy, man is called a *Saptaparna* which means a seven-petalled lotus plant which symbolises man's nature and consciousness on the three principal planes of his being: (*a*) the lotus roots in the earth drawing nutrients from the soil symbolise man's bodily dependence on physical matter; (*b*) the plant stem in the water connecting the roots at the lower end with the flower on top representing man's soul principle uniting his mortal physical existence with his immortal spiritual nature; and (*c*) the flower turning to face the sun standing for man's immortal spiritual principle.

The Rosicrucian movement sprouted in the early seventeenth century still exerts a powerful fascination and influence today. The Fraternity of the Rosy Cross as founded on the legend of the German Christian Rosenkreutz is, in its modern guise, the Fellowship of the Rosy Cross founded in 1915 by the American-born British poet and scholarly mystic Arthur Edward Waite (1857–1942). The rose, like the lotus, its Eastern equivalent, represents spiritual unfoldment and attainment. Their manifestos comprised cryptic texts depicting universal wisdom and a new philosophy based on alchemy, esoteric wisdom and religious and spiritual canons. Given their archetypal image of the rose and cross, the recovery of lost knowledge and the idea of a hidden brotherhood working for the regeneration of mankind, it is easy to understand the widespread wave of enthusiasm from the cognoscenti over much of Europe at the time.

The Qabbalah is the hidden wisdom of the Hebrew rabbis of the Middle Ages, in turn derived from older Chaldean secret doctrines concerning cosmogony. The word means 'to receive', or 'to take over', signifying something which is handed down to selected individuals by word of mouth. The Qabbalah, therefore, is essentially the theosophy of the Jews. It supplies the key without which the spiritual mysteries of both the Old and the New Testament must ever remain unsolved. The *Zohar* is generally regarded as the original and main book of the Qabbalah.[21]

The Tree of Life in the Qabbalah, portrays the ten emanations or attributes of deity, known in Hebrew as 'Sephiroth' collectively and 'Sephirah' individually. For this reason, the Sephirothic Tree is sometimes depicted as a human body, the archetype of man—a microcosm of the universe. It is important to keep in mind the idea that the Sephiroth are all linked together and there is constant, conscious interaction between them, typifying the interaction present between all grades of beings in the manifested universe.[22] Note, however, that according to Qabbalistic cosmogony (and indeed esoteric cosmogonies of all cultures and epochs), the birth of the physical universe was the last stage, not the first stage, in emanational emergence from eternity. Creation or emanation has a much wider meaning for Qabbalists than it does for materialistic scientists.

The Law of Analogy

In her esoteric instructions, Blavatsky's students were told to 'reduce everything to terms of consciousness'. (This is in stark contrast to the approach of establishment science to reduce consciousness to terms of matter in motion and its interactions.) It is not difficult to see the truth of this. Consider first man, the microcosm. It is obvious that every *external* motion, act or gesture, whether voluntary or mechanical, organic or mental, is produced and preceded by *internal* thought, feeling, will or volition. Sense experiences are really mental experiences, the movement of consciousness.

Now, applying the same principle on a larger scale to the macrocosm, just as no outward motion or change in man's external body can take place unless provoked by an inward impulse, so with the external cosmos. The whole cosmos is guided, controlled, and animated by an almost endless series of hierarchies of conscious and animate beings, each having a mission to perform, and who—whatever name we give to them—like Dhyan-Chohans in the East or Archangels in the West—are 'messengers' in the sense only that they are the agents of karmic and cosmic laws.

Blavatsky advises: '*Analogy* is the guiding law in nature, the only true Ariadne's thread that can lead us, through the inextricable paths of her domain [the labyrinth of confusion, sensuality, and danger], toward her primal and final mysteries.'[23] Analogy, therefore, is the surest guide to the comprehension of the occult teachings because, 'Everything in the Universe follows analogy. "As above, so below"; man is the microcosm of the Universe. That which takes place on the spiritual plane repeats itself on the Cosmic plane. Concretion follows the lines of abstraction; corresponding to the highest must be the lowest; the material to the spiritual.'[24]

Blavatsky adds that, 'The Universe is worked and *guided* from *within outwards*. As in heaven so on earth; and man—the microcosm and miniature copy of the macrocosm—is the living witness to this Universal Law and to the mode of its action.'[25]

But it must be remembered, continues Blavatsky, 'that the study of Occultism proceeds [by the Platonic method] from Universals to Particulars, and not the reverse, as [almost universally] accepted by Science'. 'As Plato was an Initiate, he very naturally used the former [deductive] method, while Aristotle, never having been initiated, scoffed at his master, and, elaborating a system of his own, left it as an heirloom to be adopted and improved by [Roger] Bacon.'[26]

Correspondences therefore exist between the principles of physical nature and the human principles described earlier, namely, the four Elements of physical nature and the Lower Quaternary of man, each principle of which may be regarded as a 'human element', so to speak. A living organism is composed physically of approximately seventy percent water and the remainder comprises organic molecules, the key elements of which are hydrogen, nitrogen, oxygen, and carbon. At the base of all organic molecules is the element carbon whose versatility enables the formation of all biomolecular structures. In most organic molecules, a linked chain of carbon atoms forms a backbone to which other atoms attach. Covalent bonds (chemical bonds between two non-metal atoms) between carbon and hydrogen atoms form the basis of the simplest organic molecules (such as methane gas). Other key elements are oxygen and nitrogen—both play crucial roles in life processes by modifying the structure and chemical properties of organic molecules. Other elements, like sulphur and phosphorus, have important roles but the four elements just stated are the crucial ones.

Esoteric Cosmogony – Kosmic and Terrestrial Planes

We now explore how the planes upon which the substance of man's various bodies are derived are the mirror of the corresponding planes of cosmos and how the seven principles comprising the constitution of man, the microcosm, are related to corresponding principles in cosmos, the macrocosm.

One of the finest attributes of the Hermetic Axiom is that immense complexity can be concentrated or reduced to elegant simplicity through the law of analogy. Materialistic science attempts to comprehend the universe by reducing physical matter to elementary *particles*. Esoteric science fathoms the cosmos through fundamental or elementary *principles* and occult science views material particles as the highly attenuated state of 'Spiritual particles', i.e., super sensuous matter existing in a state of primeval differentiation.[27]

The immensely complex inner and outer nature of man is analogous to that of the total universal being. This applies not only to man but also to every other entity. Everything reflects the whole, actually or potentially, to a greater or lesser degree dependent on the 'closeness' of each entity to its source. There is no outward resemblance between the physical body of man and the physical solar system or the cosmos. But if we search behind outer appearances and examine the principles, potentialities, and tendencies in the outer forms, it is always possible to discern a remarkable similarity between man the microcosm and Cosmos, the Macrocosm.

The relatively new concept of the multiverse or meta-universe was completely unknown during the nineteenth century when the Theosophical doctrines were formulated with the main objective of presenting the ageless wisdom and occult sciences through the idiom of the latest science of the day. 'Multiverse' is an umbrella term in physical cosmology for the universe in which we live, *plus the hypothetical set of all other possible universes*. It has resulted from recent developments in cosmology and particle physics, that have led to the remarkable realisation that our universe—rather than being unique—could be just one of many universes.

Together, these universes comprise everything that exists: the entirety of space, time, matter, energy, and the laws and constants that govern them. The various universes within the overall multiverse are called by names such as 'parallel universes', 'other universes' or 'alternative universes'. Since the physical constants may be different in other universes, the fine-tunings which appear to be necessary for the emergence of life may possibly be explainable. Nevertheless, many physicists remain uncomfortable with the multiverse hypothesis since it is highly speculative and always likely to be unamenable to experimental testing (for obvious reasons).[28] Notwithstanding the above, primordial gravitational waves in the universe or ripples in space-time, which also appear in Einstein's theory of general relativity, arguably lend support to the recent theory that the universe is constantly giving birth to smaller 'pocket' universes within an ever expanding multiverse.

Further support for the multiverse theory was provided in a paper by Stephen Hawking (1942–2018) a few weeks before his death in 2018. It reportedly provides mathematical evidence for the theory that there have been multiple 'Big Bangs', of which our own universe is just one such event offering the possibility of universes existing well beyond our own with completely unknown galaxies, stars, and planets.[29]

Kosmos and Cosmos

In general, the term 'cosmos' written with a lower case *c* is used in the commonly accepted meaning attaching to the word: our solar system comprising the sun and the celestial objects that orbit it, either directly (like the planets) or indirectly (like the planetary moons). Cosmos, written with an upper-case C refers to both the objective and subjective aspects of our solar system.

However, 'Kosmos' (written with an upper-case *K*) denotes the numberless, infinite universes, the innumerable solar systems of the infinite Kosmos. Written with a lower-case *k*, kosmos refers to the objective aspect of Kosmos. *The term especially implies the Womb of all Universes to be.*

All the great philosophies and religions of past times, likewise all the ancient sciences, taught the fact of kosmic life: the existence of inner, invisible, intangible, but *causal realms* as the foundation and background of these innumerable systems. Thus, our physical world (like our physical body) is but the outer veil of other worlds which are inner, vital, alive, and causal and which, *in their aggregate* embody the Kosmic Life.

It must be understood that the terms 'Kosmos' and 'Universe', as used in the esoteric philosophy, signify above all else, the indwelling boundless life expressing itself through the myriad forms and entities producing the incredible variety—unity in diversity—that we see and experience around us.

As depicted in Figure 3 below, Kosmos comprises seven principal planes: six inner, subjective planes and one outer objective plane. Of these, the inner topmost three planes pertain to the Unmanifest Worlds and the remaining four to the Manifest Worlds. Each plane is stratified into seven sub-planes, each of which has seven sub-sub-planes and so on. This accounts for the infinite variety and richness in all departments of nature. Bring to mind the beautiful colours of natural scenes like a sunset, a rainbow or the countryside. It requires but three primary colours that can be combined to produce innumerable combinations, infinite subtlety and a variety of colours in the visible spectrum.

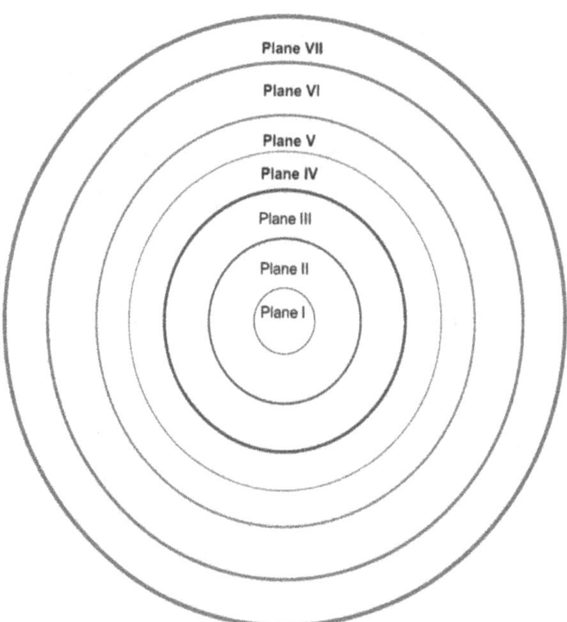

Figure 3: The Seven Planes of Kosmos – a Diagrammatic Representation

Every plane and sub-plane has several correspondences and constitutes a characteristic level or mode of consciousness spanning the Unmanifest inner, topmost three planes of the

Divine and Formless World of Spirit; and the Manifest outer four planes designated: Archetypal, Intellectual or Creative, Substantial, or Formative, and Physical or Material. These are the four lower planes of Cosmic Consciousness, the three higher planes being inaccessible to human intellect as developed at present.

Each plane has its own 'field of force' being a 'sphere' or 'world' in its own right. Therefore, beings functioning in any one of the seven worlds are quite independent of the other worlds and are able to function without interference from other worlds.

But there is an important fact to note: at any level beings of the innermost worlds, because of their loftier status, are able to penetrate to the outermost world whereas those of the outermost world are unable to contact the inner planes and must therefore function solely in their own world. This is one of the meanings of that mysterious term 'Ring Pass-Not'.

In the arcane occult cosmogony, we sometimes come across the term 'Older Wheels' which refers to worlds or globes pertaining to other solar systems as they were during previous epochs of the cosmos. Invoking, again, the Hermetic Axiom, since human rebirth exists on the smaller scale, there is no reason why this analogous principle should not apply to Cosmos on the grand scale. This is in fact the case. Countless globes evolve after a periodical Pralaya or cosmic slumber, rebuilt from older material into new forms. The previous globes disintegrate and reappear transformed and perfected for a new phase of life.

It is important to stress here that the universe is self-regulating, by virtue of divine law, there being no external 'God' acting as external law-giver or regulator.

There are seven *planes* that correspond to the seven *states* of consciousness in man. The three higher planes of Cosmic Consciousness, pertaining to the Unmanifest Worlds, are formless (*Arupa*), where form ceases to exist on the objective plane. These higher planes are revealed and explained only to initiates. The four lower planes in descending order, restated for convenience, are the Archetypal or Prototypal World, the Intellectual or Creative World, the Substantial or Formative World and the Physical or Material World—being the lowest plane, the visible cosmos.

In order to build anything, there are four basic stages involved: firstly ideation, then an intellectual (mathematical) or physical model of the proposed idea, followed by the detailed design of the 'nuts and bolts' and, finally, the finished product in physical matter conforming to the original idea.

Mainstream cosmology tries to fathom the secrets of creation by exclusive emphasis on the material process alone and ignores the first three! If a 'Big Bang' is supposed to have occurred, did it not first exist as an Idea in Divine Mind, then as Primordial Thought, followed by Divine Design and finally Divine handiwork—the universe of matter?

Humanity will evolve in the far distant future to a stage where its collective intellect will touch these higher three planes of consciousness. During our present epoch, however, it is *within* the four lower planes of kosmos that the whole drama of human evolution occurs and the sevenfold stage on which it is enacted. Of these four planes, three are super-physical or subjective, our Earth alone being on the plane of sensory perception.

It is important to repeat the cardinal occult doctrine that every celestial body, be it a planet, comet, nebula, atom, electron or quark, is a composite entity comprising (*a*) inner and invisible energies and substances and (*b*) an outer and generally visible physical vehicle.

Each of the physical globes scattered in the immense reaches of space is accompanied by six invisible 'companion globes', forming what in Theosophy is called a 'Chain of Globes'. This is the case with every sun or star, planet and moon. All have a sevenfold constitution—like man who is a miniature copy of the universe. As every entity in the universe is an inseparable part of it, what is in the whole is reflected in every part, because the part cannot contain anything that the whole does not contain. But it contains the whole only *in terms of itself* and this is why it cannot supersede or become greater than the whole.

Six of the seven globes of our Earth chain (as with other planetary schemes in general) must be envisaged as existing on the subjective or inner planes. Three are on the arc of descent towards increasing materiality known as *involution* and three on the arc of ascent towards an increasingly ethereal condition—known as *evolution*.

The globes are not to be regarded as distributed in regions of three-dimensional space but interpenetrating and occupying the same 'space'. Only one of these seven principles, the physical, is perceptible to the physical senses.

Akasha, the Astral Light, and Memory

As there are seven planes of being (six of which are non-physical), there are seven corresponding principles in man. Each of these planes is a differentiation of one primordial root substance or matter, known under a variety of names such as *Svabhavat* in the East and Æther or Akasha in the West. These terms mean 'essential nature' or the spirit and essence which is behind 'world substance'.[30] It is important to note that there is just one substance—Akasha—in progressively lower gradations of *itself*.

As a general principle, a body is constituted of the material of the plane that it inhabits. So, for example, a car made of steel and alloys is fabricated from iron ore and other metallic elements mined from the Earth. Likewise, the physical body uses terrestrial material drawn from that plane. So, also, the subtle bodies (like the astral body) draw upon the appropriate grade of subtle material from the non-physical planes composed of finer substance.

Accordingly, the physical body of man is constituted of the physical substance of the lowest, i.e., seventh plane of kosmos—our physical plane of existence. The next lowest, or the first plane above the physical, is the astral plane constituted of 'astral substance', which has its own vitality, energy, and dynamism. Its functions correspond to the non-physical astral body of man. But just as there are physical processes occurring on the Earth that are not directly related to the physical body of man, there are also aspects or 'regions' of the astral plane not directly related to man's astral body. These are collectively known as the 'Astral Light'. In both cosmic and human terms this may be regarded as the final interface between the higher subjective planes and the objective physical plane.

The Astral Light is an indispensable element to a study of consciousness, the more so because therein lies the clue to the function and operation of memory. For it is in the Astral Light that the memory of man and nature is 'stored' indelibly and virtually forever—and not, in the case of man, in his fleshly brain which acts as a retrieving mechanism. And this explains the ability of highly trained occultists, like Blavatsky, to access nature's memory banks and 'bring down' facts and details long forgotten. (It is a historically verifiable fact that Blavatsky had a library of around thirty books. Yet she was able to quote extensively and accurately from well over twenty thousand books and manuscripts, archaic and modern, providing verifiable page numbers and references.[31])

It remains with man to awaken and attune to the three higher states latent in himself in order to contact the three higher planes of kosmos. It is a secret reserved for the very few. Therefore, the constitution of man mirrors the whole of which he forms a part. These planes and their sub-planes constitute a series of vibratory scales that encompass the whole gamut of possible being. The higher the rate of vibration of any manifest being, the closer it is to its source.

Each individual can contact higher planes by means of ever-increasing purity of motive and refinement of character, turning the human into what is popularly termed the superhuman state but at non-physical levels. Such advanced intelligences can, if needed for a specific purpose of service, assume a human *form*. As such, they represent the highest orders of consciousness to be attained and indeed such higher beings have guided and provided the impulse to nascent humanity in its evolutionary development.[32] This is the basis of ideas like, 'God was an astronaut' or that 'advanced aliens in spacesuits from outer space once visited the Earth', which are not total rubbish, but the dregs of a lofty occult doctrine grossly distorted through the lens of anthropocentric phantasies, unchecked by intellect and reason.

Karmic Implications

According to the Mystery teachings, not all the spiritual nature of man incarnates in matter in one lifetime. In daily life we can sense the truth of this in terms of the gulf between our potentiality and our capacity: the common feeling that we could accomplish and create far more were it not for the limitations of our physical energy and time.

We also sense this fractional incarnation of our greater Self in the operation of karma. Stated briefly, the four types of karma are: (*a*) *sanchita,* the totality of karma from all lifetimes; (*b*) *prarabdha,* the karma for a particular birth, or incarnation; (*c*) *kriyamana,* the immediate karma from present actions; and (*d*) *agami,* the returning karma from this and other lifetimes.

We can appreciate that if the karma from all lifetimes were packed into a single lifetime, the latter would be totally overwhelmed. It is hard enough to expiate immediate karma and hopefully most of present life karma over the course of a lifetime! Our incarnations are proportioned according to how much karma (both 'good' and 'bad') can productively be processed in one physical life. Therefore, according to the spiritual laws of conservation of matter and energy not all the spiritual nature of man incarnates in matter in a single incarnation.

Accordingly, at birth, only a third part of the divine nature of man temporarily dissociates itself from its own immortality and takes upon itself the dream of physical birth and existence, animating with its own celestial enthusiasm a vehicle composed of material elements part of and bound to the material sphere.[33] At death this incarnated part awakens from the dream of physical existence and reunites itself once more with its eternal condition. This periodical descent of spirit into matter is termed the wheel of life and death.

This is the secret meaning behind the affirmations of the ancient philosophers when they declared that the rational part of man's soul never entered wholly into the man, but only overshadowed him, more or less, through the irrational spiritual soul.

Origins of Humanity

The origin and development of the human species is necessarily a complex affair but ultimately it comes down to a three-staged emanational descent.

1. For its moral, psychic, and spiritual nature mankind is indebted to a group of divine Beings known as the Watchers, or Regents.
2. Mankind, in its first prototypal shadowy form is the offspring of the *Dhyani-Chohans* (the collective hosts of spiritual beings equivalent to the Elōhīm, the Angelic Hosts of Christianity). These are celestial beings charged with the supervision of the Kosmos by way of administering and enacting Divine Laws. They are 'celestial' in the sense of being at a stage on the ladder of life superior to the human kingdom. This is because they have evolved through the human stage in far past aeons.

3. In its qualitative and physical aspect, humankind is the direct progeny of the lowest grade of *Dhyanis* or Spirits of the Earth.

Collectively, then, men are the handiwork of hosts of various spirits—the hierarchy of active cosmic agents on the various planes. Man is not, nor could he ever be, the complete or sole product of the 'Lord God' as some religious fundamentalists would proclaim—any more than a temple can physically materialise from the vision of the architect without a definite plan actuated by builders and craftsmen. But man is the child of the Elohim—the minor created gods—so arbitrarily changed into the singular masculine gender.

The first Dhyanis, commissioned to 'create' man in their image, could only throw off their shadows like a delicate model for the nature spirits of matter to work upon. Man is, beyond any doubt, formed physically out of the dust of the Earth, but his creators and fashioners are many.

Accordingly, man is composed of countless myriads of lives. Science perceives this truth at the biological level, being well aware of the countless bacteria and viruses and other infinitesimals in the human body, their role in health and as occasional and abnormal visitors to which diseases are attributed. Occultism—which discerns a life in every atom and molecule, whether in a mineral or human body, in air, fire or water, in bacterium or virus—affirms that our whole body is built of such lives.[34]

NOTES

1. *SD*-I. 'Summing Up', 274.
2. Philip Carr-Gomm, Druid Mysteries: Ancient Wisdom for the 21st Century (UK: Rider, 2002).
3. *STA*, op. cit., CXXXI.
4. *STA*, op. cit., CXXX.

5. *STA*, op. cit., CXXXII.
6. Ernst Lehrs, Man or Matter: Introduction to a spiritual understanding of nature on the basis of Goethe's method of training observation and thought, rev. and enl., Nick Thomas and Peter Bortoft (3rd edn, London: Rudolf Steiner Press, 1958), 243.
7. Alexander Wynne, 'The Oral Transmission of Early Buddhist Literature', Journal of the International Association of Buddhist Studies, 27/1 (2004), 97–127.
8. *STA*, 'The Tabernacle in the Wilderness', CXXXV.
9. C. C. W. Taylor and Mi-Kyoung Lee, 'The Sophists: 1. Protagoras', The Stanford Encyclopedia of Philosophy (18 August 2020) <https://plato.stanford.edu/entries/sophists> accessed 17 October 2020. See also D. Bostock, Plato's Theaetetus 151e (Oxford: Clarendon Press, 1988).
10. This section and the next taken from *STA*, 'The Human Body in Symbolism', LXXIII.
11. *STA*, op. cit., LXXIV.
12. *STA*, 'The Human Body in Symbolism', LXXIV.
13. 'The Brain: An extract on occult physiology from The Collected Writings of H. P. Blavatsky', Theosophy in New Zealand (June 2005), from *CW*-XII, 'E. S. Instruction No V', 697–9.
14. Craig Hamilton, 'Is God All in Your Head? Inside science's quest to solve the mystery of consciousness', What Is Enlightenment? Magazine Reprint Series, 29 (June–August 2005), 95.
15. Rollin McCraty, Science of the Heart, ii: 'Technical Report – Exploring the Role of the Heart in Human Performance: An overview of research conducted by the HeartMath Institute' (Boulder Creek, California: HeartMath Institute, February 2016), 36.
16. Jay Pasricha, 'The Gut-Brain Connection', John Hopkins Medicine <https://www.hopkinsmedi- cine.org/health/

wellness-and-prevention/the-brain-gut-connection> accessed 7 July 2020.

17. Paul Pearsall, The Heart's Code (New York: Broadway Books, 1998).
18. Asok K. Mukhopadhyay and Jay Relan, 'The Deep Science of Neuro-Cardiology', Annals of Psychiatry and Clinical Neuroscience, 3/2 (June 2020), 1.
19. 'Teilhard de Chardin', Xavier University <https://www.xavier.edu/jesuitresource/online- resources/quote-archive1/spiritual-awareness-quotes> accessed 7 July 2020.
20. Albert Einstein', reddit <https://www.reddit.com/r/quotes/comments/clbjne/the_release_of_ atomic_power_has_changed> accessed 27 June 2020, quoted in Asok K Mukhopadhyay and Jay Relan, loc. cit.
21. See G. de Purucker, Occult Glossary: A compendium of Oriental and Theosophical Terms (1933; Pasadena, California: Theosophical University Press, 1972), 137, 138.
22. See Geoffrey Barborka, The Divine Plan (2nd edn, rev. and enl., Adyar, Madras: Theosophical Publishing House, 1964; repr. 1980), 50.
23. *SD*-II, 'Stanza VI: The Evolution of the Sweat Born', 153.
24. *SD*-I, 'Explanations Concerning the Globes and the Monads', 177.
25. *SD*-I, 'Summing Up – The Pith and Marrow of Occultism', 274.
26. See *CW*-XII, 'E. S. Instruction No. III', 599.
27. *SD*-I, 'Stanza VI: Our World, Its Growth and Development – Stanza VI Continued', 200–1.
28. B. J. Carr (ed.), Universe or Multiverse? (New York: Cambridge University Press, 2007).
29. S. W. Hawking and Thomas Hertog, 'A Smooth Exit from Eternal Inflation?' Journal of High Energy Physics, 147 (27 April 2018).

30. See Ervin Lászlő, Science and the Akashic Field: An integral theory of everything (Rochester, Vermont: Inner Traditions, 2007).
31. William Kingsland, The Real H. P. Blavatsky: A study in Theosophy, and a memoir of a great soul (London: Theosophical Publishing House, 1985), 226–7. See also: Constance Wachtmeister, Reminiscences of H.P. Blavatsky and 'The Secret Doctrine' (London: Theosophical Publishing House 1976); Geoffrey Barborka, H. P. Blavatsky, and Tibetan Tulku (Adyar, Madras: Theosophical Publishing House, 1986).
32. One of the finest and most accessible books based on a commentary on archaic records in The Secret Doctrine is by Geoffrey Barborka, Peopling of the Earth (Wheaton, Illinois: Theosophical Publishing House, 1975).
33. *STA*, 'The Human Body in Symbolism', LXXVI.
34. Adapted and rephrased from: *SD*-I, 'Stanza VII—Continued: Spirit Falling into Matter', 224, 225; Geoffrey Barborka, The Divine Plan, 59, 60, 61.

Part Five

A New Science of Consciousness

Between the shores of the ego and the ocean of the soul there are many planes of consciousness.

Revd.Professor Stephen Wright[1]

It is crucial to acknowledge the primacy of consciousness. Because every idea that we may have, all our dreams and aspirations, joys and sorrows, in fact, all that we know is experienced by, in and through consciousness. To discover a neural correlate of consciousness and thereby solve the problem of the experience of sensation are currently white-hot topics of scientific debate and neuro-scientific research. What is the neural link between, say, external auditory perception or even just the thought of music and the internal, subjective experience?'

Neuroscience and cognitive science have not, by their own admission, come up with a satisfactory answer even after decades of massive research. However much they may have identified specific regions in the brain and associated cerebral mechanisms related to vision and the other senses, the question of how that transforms to sensation, as an experience, has doggedly eluded these scientific disciplines.

So let us give occult science a fair hearing. The detail is highly complex and need not concern us here as there are erudite expositions in occult literature,[2] but the process can be outlined. Even a hazy understanding is preferable to the cul-de-sac of materialistic hypotheses.

A science of consciousness must explain the exact relationship between subjective mental states and objectively measurable brain states—the relationship between the conscious mind and the electro-chemical interactions in the body.

Why is neuroscience currently experiencing such difficulties and virtually intractable problems in finding a neural correlate of consciousness? There are fundamentally three reasons:

1. The erroneous assumptions, and ensuing perceptions, about the relation between the senses and the mind.
2. The limited outlook on the whole constitution of man.
3. The failure to distinguish consciousness from its vehicles of expression.

Regarding the first, mainstream neuroscience adamantly maintains that the senses are the basic reality and that the mind, being a product of the senses, is therefore subordinate to the sense organs. In fact, it is the other way around in occult psychology.

The second reason—science's limited outlook—is due to its viewpoint constrained to the physical domain only, by virtue of its determined resistance to blend physics with metaphysics so as to investigate the whole constitution of man, namely, the physical frame alongside his non-physical counterparts, that is to say, the body and its animating principles. In contrast, occult science has investigated the whole length and breadth of the constitution and function of man and is therefore qualified to pronounce on this physical-to-mental conundrum for science.

The third reason stems from the unwarranted assumption that the vehicle, i.e., the brain, is the mechanism considered to produce thought and consciousness is an emergent property of material processes in the brain.

So even if, hypothetically, a satisfactory neural correlate were found, it could not in any way explain the nature of sensation or experience.

Space, Time, and the Role of Maya

Creation, strictly speaking, is a misnomer. In the truest sense, nothing is ever created. Creation is really a process of emanational emergence, which literally means 'to flow out and bring forth'—from divine consciousness or the unmanifest—to its manifested appearances on various planes of descent culminating in the physical. In summary:

1. The basis of the universe, or what we call the Ultimate Reality, is Divine (Supreme) Consciousness.
2. The mind itself is a phenomenon in the basic medium of consciousness, being a modification of consciousness presenting as energetic intelligence, or intelligent energy.
3. Matter, or crystallised energy, is of the essential nature of mind, which explains why matter is never inert but displays intelligent action at its own level, as shown by chemistry. Hence the physical universe of matter is, in essence, a purely mental phenomenon without any intrinsic physical-material basis at all, which quantum science has conclusively shown.

It is as well to clarify that emanation is not the same as evolution—nor creation. The word emanation derives from the Latin *emanatus* meaning 'to flow out', namely, 'to pour forth', or 'to pour out of'—the mode by which all things are derived from REALITY, the first principle. So, all derived or secondary things proceed or flow from the more primary. Evolution means the unwrapping, or unfolding, of latent powers within the entity itself.

It is the impact and mirroring of Universal Divine Mind on our individual minds that are, in most cases, responsible for our individual mental worlds. It is because the world process has its origin in Divine Ideation that, quite naturally, each individual will receive his own impression of the process according to his position in space and time. This means that those who occupy the same position will receive similar impressions. Such positioning of course means mental positioning—mental co-ordinates—which accounts for the similarity of experiences of different individuals, coloured and conditioned by the development and sensitivity of their own individual minds. Throughout history there have been cases of minds attuned to a common high purpose that will produce works of art or science having a common exalted characteristic.

What Mainstream Neuroscience Proposes

In *Living in a Mindful Universe,* the Harvard neurosurgeon Eben Alexander, mentioned earlier, provides a useful discussion on why materialism is at a loss to explain the above question. Drawing deeply on his practice of decades of neurosurgery and, crucially, on his own near-death experience, he states that 'materialist science [...] posits that the brain creates consciousness out of purely physical matter—that there is nothing else.' Accordingly, on the basis of this unproven assumption 'it asserts that everything we have ever experienced—every beautiful sunset, every gorgeous symphony, every hug from our child, every experience of falling in love—is merely the electrochemical flickering of around a hundred billion neurons in a three-pound gelatinous mass, sitting in a warm dark bath inside our head.'

However, the problem with 'the materialist brain-creates-consciousness model, is that not even the world's top experts on the brain have even the remotest idea *how* [*sic*] the brain could create consciousness,' other than just taking it as read that it does.[3] For example in an article entitled *'Brain Reactions'*, we are explicitly informed that it is in 'our cerebral cortex, where consciousness arises.'[4]

For the moment, we may confidently assert that if the brain does not produce consciousness any more than it actually produces sound waves when we hear music, we need no more convincing that, as Alexander says, 'materialist science as a foundation for comprehending our reality is at a dead end'.

What Occult Science Affirms

The transmutation of physical nervous impulses to mental sensations in the human being is brought about through the agency of one of the divine forces—prana—serving as the vehicle of the vital force. Prana may be thought of as a special kind of transmuting substance—a compound of physical matter and mind. Being on the interface between the two it therefore enables mind to affect matter and matter to affect mind. The simile with a chemical catalyst is useful. A catalyst is a substance that actuates or accelerates a chemical reaction while undergoing no permanent change in itself, nor is it consumed by the reaction. Therefore, it can be recovered chemically unchanged at the end of the reaction. Note that the action goes both ways: matter and mind can affect one another.

The reference to prana should not be taken as a literal concept of matter, as in the case of atoms and molecules, since science has explicitly proved that such physical entities are in reality modes of motion or 'bottled-up waves'. There is no logical reason why this *vibrational or wave nature of matter should not extend from physical matter to the subtlest grades of supra-physical realms*. Paradoxically, then, for the scientist, but quite logically for the occultist, it is waves affecting mind or non-physical substance affecting mental substance. Like affects like, and there is no dichotomy between matter and mind *when the essentially mental or mind-based nature of the former is grasped*.

What Enlightened Science Intimates

The role of prana as the agent of the vital force is central to the whole question of how physical neurology can result in an internal experience. It is worth asking an open question as to whether any quantum processes could explain the 'hard' problem of consciousness. This is now a burgeoning field of enquiry amongst several enlightened scientists of international acclaim like Roger Penrose (*b.*1931) and Stuart Hameroff (*b.*1947), who have long argued that consciousness is the result of quantum gravity effects in microtubules (tubular polymers of tubulin) which they dub 'orchestrated objective reduction'.

There is no suggestion, however, that quantum theory exclusively can explain consciousness. Rather consciousness is a key factor in a complete understanding of quantum phenomena.

Problems in Neuroscience Resolved by Occult Science

Occult science leads us, step-by-step, by its own inexorable logic and self-consistency, to the inference that behind the apparently physical world, governed by the physical laws known to science, is an underlying mental world containing the noumenal archetypes, principles, modes of motion, and images—all mathematically related and harmonised into an organic whole.

Despite the truly infinite variety and levels of existence in manifestation, there is only one Ultimate Reality from which every variety is derived and in which all aspects of manifestation, without exception, are contained or subsumed. The Absolute is the root and fountainhead of the entire Cosmos, seen and unseen, manifest and unmanifest. It is Divine (Supreme) Consciousness.

The objective universe, cognised through the five sense organs, is nothing but an expression of the same Reality which functions as the cogniser and 'sees' this objective universe through the three eyes: two of the eyes are the normal eyes used for instrumental perception while the Third Eye represents non-instrumental perception.[5]

Awakening Latent Powers

The senses are to be regarded as the extended instruments of mind and subordinate to the latter, instead of the conventional scientific view that the senses are the basic reality and mind the product of sensations received through the sense organs. Additionally, there are other consciousness-centres and glands in the human body that are particularly connected with the awakening of the dormant latent forces in man. Two such main centres are the pituitary and pineal glands.

The whole human being, but especially the heart, brain, and generative system as main centres of consciousness, may be regarded as a divine instrument that transforms Universal Consciousness into its physical substrate.

The entire corpus of the Perennial Philosophy affirms that the conditioned universe is an aspect of the ONE REALITY: 'The Eternity of the Universe *in toto* as a boundless plane; periodically "the playground of numberless Universes incessantly manifesting and disappearing," called "the manifesting stars," and the "sparks of Eternity." "The Eternity of the Pilgrim" is like a wink of the Eye of Self-Existence [...]. "The appearance and disappearance of Worlds is like a regular tidal ebb, flux and reflux." '[6]

'Pilgrim' is the monad during its cycle of incarnations. It is the only immortal and eternal principle in man, being an indivisible part of the integral whole—the Universal Spirit, from which it emanates, and into which it is absorbed at the end of a cycle.

Occultism affirms that the entire Universe, in its visible and invisible worlds, is the periodical manifestation of the Unmanifest Reality, which: (*a*) emanates 'numberless Universes'; (*b*) expresses itself through it for a limited period of time; and then (*c*) withdraws it into itself—the whole cycle being repeated numberless times.

The universality of that law of periodicity is not difficult to perceive—day and night, life and death or sleeping and waking. It is easy to comprehend that in it we see one of the absolutely fundamental laws of the universe.

However, in order that the Unmanifest Reality may manifest, It must first create, or rather emanate, the raw material which can be worked-up into suitable forms or vehicles for Its self-expression through them according to Its sensitivity and responsiveness. The three well-defined stages involved in manifestation are creation of the raw material, evolution of the vehicles, and expression of REALITY through the vehicles and unfoldment of their infinite potentialities.

Fohat, Prana, and Kundalini – Three Divine Forces Still Unknown to Science

There are three different types of energy poured out from the sun (strictly speaking, the Solar Logos of which the physical sun is its outer 'clothing') and appropriated by the human body for its various functions.

Despite its minute and detailed knowledge about the body, science only acknowledges one of the three divine forces operating in it—electricity. In fact, The Royal Institution of Great Britain recently (March 2018) hosted a discussion on 'The Electricity of Life', exploring how our body creates electricity, how it was discovered, and what clues our understanding of electricity can give to help cure disease—although science, as yet, would hardly want to hear about electricity derived from Brahma, and still less its correspondence with *Fohat*, the creative force of Brahma.

All material changes in the human body are brought about by electricity and other related forces (such as electromagnetism), and their associated organising fields known to science. These are related to the form or material side of the body and nature.

Science has had suspicions about another force referred to vaguely as 'vitalism' and connection with prana, or Chi, a refined medium: necessary for the production of form; by means of which all motion is conveyed; through which the outer living world communicates with the living man; and when absent, leaves the body actually dead (see later for more detail). But any such notions of a vital principle were effectively anathematised by the principal advocates of molecular biology who dogmatise that life is nothing other than physical chemistry.

Prana is related to the life side of the body and nature and is responsible for all life processes in the body to make it a living organism rather than an insentient aggregate of forces and matter. Most significantly, prana lies at the basis of sensation and without its association with a sense organ the vibrations received by that organ would remain ineffective and would not be converted into sensation. Hence, the endeavour by neuroscience to discover a neural correlate of consciousness has met with only partial and limited success and will never attain fruition until full notice is taken of vitalism and the role of prana.

The third divine force *kundalini* is unknown to science and is found only in the human being and not in any other living organisms. The reason is complex. Kundalini is intimately connected with the functioning and unfoldment of consciousness in the human body. Again, science has not unlocked, and never will unlock, the full mystery of consciousness unless it understands the role and function of this force.

It should never be imagined that during the lifetime of a human being the three forces act sequentially with only one force active at any one time. All three forces are always in action. What varies is the predominance of one force over the others.

Vitalism – and Its Death by Science

An outspoken statement against vitalism (defined below) comes from Sir John Maddox (mentioned earlier) in his lecture to the British Humanist Society: 'Life is chemistry. And vitalism is dead. That's wisdom in a true sense, and it's invaluable.'[7] This is just one example of the opinions of the vast majority of world-famous scientists who now claim the 'death of vitalism' on the basis of the joint claim by Francis Crick and James Watson that 'the discovery of DNA's double-helix structure was a major blow to the vitalist approach and gave momentum to the reductionist field of molecular biology.'[8]

If vitalism is now supposed to be 'dead' it must once have been living, in fact is still living. Moreover, scientific opinion is by no means united on this subject and, as in everything else, there are a few men of science whose views veer towards vitalism. So despite the outright denial of a vital principle by their peers in biology and physiology, these enlightened scientists have seen through the limitations of the *exclusively* 'reductionist field of molecular biology' towards an insight that, ' "We human beings" experiencing our joys and our sorrows, our memories and our ambitions, our sense of personal identity and free will, *cannot possibly* be *just* the behaviour of a vast assembly of nerve cells and their associated molecules.'[9]

Therefore, there has to be a distinct principle independent of the organism but of a substance existing in a state still unknown to orthodox science. *Life for these few legendary scientists is something more than the mere interaction of molecules and atoms.* They maintain that without a vital principle, no molecular combinations could ever have resulted in a living organism, least of all in the so-called 'inorganic' matter of our physical plane of consciousness.

Winding the clock back to the late nineteenth and early twentieth century, there have been many scientists who have postulated organising fields and vitalism.

The eminent British physician Sir Benjamin Ward Richardson FRS (1828–1896) called this vital principle 'nervous ether' emanating from the sun. He asserted that between all molecules 'there exists a refined subtle medium, vaporous or gaseous, which holds the molecules in a condition for motion upon each other, and for arrangement and rearrangement of form; a medium by and through which all motion is conveyed; by and through which the one organ or part of the body is held in communion with the other parts and by and through which the outer living world communicates with the living man: a medium which, being present, enables the phenomena of life to be demonstrated, and which, being universally absent, leaves the body actually dead...'[10]

A 'new flood of light' is certainly thrown on the wisdom of ancient and mediæval occultism and its votaries, for we have an important scientific corroboration for a fundamental tenet of occultism—that (*a*) the sun is the store-house of life-force or vital force and (*b*) that from its mysterious depths come forth life currents which thrill through space as well as through the organisms of every living thing on Earth. Occultists refer to these currents as prana, our 'life-fluid'.

From all of the above, is it not disappointing for the occultist to hear Nobel Prize laureates in science talking glibly of 'chemical energy' or, to wit, 'life as chemistry' and denouncing the parent vital energy as an exploded superstition?

The importance of prana or chi is recognised worldwide in connection with health and equanimity. *Pranayama* is one of the ancient Indian yogic disciplines for achieving that purpose. The word is translated in yoga systems as the life-force, vital air, vital energy, and life energy. Its health-giving benefits include clearing physical and emotional obstacles in the body, concentration and mental clarity by freeing the breath and so promoting the unrestricted flow of prana.[11]

Similar to *pranayama,* but using quite different techniques, is the ancient Chinese practice of *Qigong.*[12] *Qi,* or *chi* is life energy. All Qigong practices are intended to cultivate and balance chi through co-ordinated body posture and movement, moving meditation, regulated slow flowing movement, deep rhythmic breathing and a calm state of mind. It is practised variously for exercise and relaxation, preventive medicine and self-healing, alternative medicine, meditation and self-cultivation. According to Taoist, Buddhist and Confucian philosophy, Qigong allows access to higher realms of awareness, awakens the true nature of a person, and helps develop human potential.

The Pituitary Body, Pineal Gland, and the Third Eye

As mentioned earlier, occult forces are latent in two principal human organs: the pituitary body and the pineal gland These are psycho-neural transducers of consciousness, dormant under normal conditions for the vast majority of people by virtue of a protective veil, but which can be aroused under special circumstances. Our objective is certainly not to divulge guarded secrets (even if we knew about them), but to show the limitless—truly unlimited—powers latent in man.

About the size of a grain of rice and located deep in the centre of the brain, the pineal gland was once known as the 'Third Eye'. The precise physiological function of the pineal gland is still something of a mystery to medical science. However, it is known that it secretes a single hormone, melatonin, which serves two principal functions: to control the circadian rhythm of the body (24-hour biological cycle characterised by sleep-wake patterns) and to regulate certain reproductive hormones.

The pituitary gland is a pea-sized structure located at the base of the brain, just below the hypothalamus, to which it is attached via nerve fibres. It is sometimes known as the 'master gland' of the endocrine system since its physiological function is to produce critical hormones that regulate the functions of other endocrine glands and control various bodily functions.

As well as the Third Eye, the pineal gland is also known as the 'Eye of Horus' in Egypt, or, in Hindu esotericism, as the 'Deva Eye', while in the esoteric Buddhism of Tibet and the Trans-Himalayan regions it is referred to as the 'Eye of Dangma'.[13] What are all these various terms pointing to? In order to understand this, it must be borne in mind that every phase of evolution reflects a profound intelligence and wisdom whereby the inner forces of consciousness work through various psycho-physical centres or chakras, among the highest of which is the pineal gland. These centres continue to develop as man evolves spiritually.

However, evolution in the fullest sense also includes previous cycles of involution. Involution means the generation of different forms of increasing concretion as the vehicles for expressing essential types and faculties. It is concerned with developing self-expression through limitation, hence it has a constraining effect on consciousness—in popular terms, the imprisonment of spirit in matter. Evolution, on the other hand, refers to an expansion of consciousness by transcending the limitations of form—the release of spirit from matter through conscious knowledge. Science considers just evolution (and only at the physical level) having practically ignored the involutionary cycle and so by restricting itself to less than half of the overall process, can be in no position to pronounce on a complete theory of evolution.

When we think of the 'Third Eye' several images come to mind from the universal myths and traditions of the world over. We may recall stories about the one-eyed giant Cyclops of Greek myth or the mystical Eye of Shiva, representing intuition or direct cosmic vision. These mythical traditions also feature early races of one-eyed giants and titans said to have lived long ages ago. The timeless wisdom of this ancient lore provides vital knowledge about how mankind came into existence on Earth.

The timeless wisdom has informed us that in the hoary past man did not inhabit his present 'coat of skin' but had a body of much more ethereal substance. Not being immersed in physicality at that time, his spiritual and psychic faculties were operative in proportion to the dormancy of his physical faculties. Then over the long course of involution—the progressive burial of spirit in matter—his spiritual and psychic nature diminished in proportion to the ascendency of his physical form.

His deepest immersion in matter occurred during the latter stages of the Fourth (Atlantean) Root Race. From then onwards began the slow evolutionary cycle of the release of spirit from matter. After aeons man will arrive at the same point of his original spirituality but only this time in full self-consciousness as opposed to the instinctive state experienced earlier. Man will cycle backwards to his earlier state of spirituality by cycling forward in his evolutionary development.[14]

During those early pre-historical epochs man had one eye, the eye of spiritual sight (located at the back of the head and not between the two physical eyes in the forehead). Then when he developed two eyes, the Third Eye—the Deva Eye— gradually atrophied and became what is now called the pineal gland.

However, this organ can, by specific techniques, be awakened to confer the spiritual insight that was once its role. As Blavatsky avers, the pineal gland is 'the very key to the highest and divinest consciousness in man—his omniscient, spiritual and all-embracing mind'. And moreover: 'This seemingly useless appendage [to allopathic medicine] is the pendulum which, once the clock-work of the *inner* man is wound up, carries the spiritual vision of the EGO [the Higher Self] to the highest planes of perception, where the horizon open before it becomes almost infinite...'[15]

Even nowadays the spiritual function of the pineal gland is not entirely defunct. We recognise its activity whenever we have a flash of spontaneous understanding about a situation, ranging from a hunch to clairvoyant abilities and intuitive awareness. How well it functions in each of us depends on how much we foster our spiritual capacities.

Practical occultism, founded on the esoteric philosophy, recognises the direct and intimate connection between the pineal gland and the genitalia located at opposite ends of the spine. These two are creative poles and when one is positive and active, the other is negative and passive in proportion. Thus, when the north pole of the spine, the pineal gland, is active it gives birth to ideas, thoughts and intuitions. When the south pole, the generative organ, is active children are procreated.

In the ordinary individual, the genitalia and pineal gland are both active by turns and therefore man is a mixture of lust and love, of passion, and compassion. We see evidence of this strong polarity in highly creative people (especially musicians, painters and performing artists) or powerful and influential public figures (like money-and-power- crazed politicians and billionaires) whose personal lives are invariably replete with diverse sexual encounters.

But in no sense whatsoever is there the least implication in this, or for that matter in any of the holy scriptures or occult doctrines, that sexuality *per se* is debasing or adverse to spirituality. This entirely mistaken, moralistic idea invariably prevails in religious or spiritual societies when the living message becomes obscured by dogma and dead ritual.

So, what is the occult function of the pituitary gland and its relation to the pineal gland? The occult wisdom affirms that the pituitary gland is connected with pure psychic visions in contrast to the pineal gland which is linked with spiritual visions. The pituitary gland is therefore the organ of the psychic plane. Pure psychic vision is caused by its molecular motion. When molecular action is set up in the pituitary, these flashes are seen and further action leads to psychic vision.

Similar motion in the pineal gland gives spiritual clairvoyance. The pineal gland is the focus of the spiritual sensorium (the apparatus of spiritual perception considered as a whole). However, the pituitary gland is only the servant of the pineal gland, its torch-bearer.

The faculty of Buddhi or wisdom-intuition exists in man but in the vast majority of cases it is not so much atrophied, as dormant. However, when there is a conjunction of Buddhi with Higher Manas (Higher Mind), Buddhi is awakened and activated, a process known as the opening of the Third Eye and represents the exercise of spiritual will.

Although the Deva Eye no longer functions for the vast majority of mankind, there was an epoch when what is now referred to as paranormal, supernatural, weird or abnormal once belonged to the *normal* senses common to all humanity of that era. These include thought transference, clairvoyance, clairaudience, and much more.

At times, the spiritual instructors of humanity have referred to what is known as 'Root Races'. The term has nothing whatsoever to do with racial or ethnic background—least of all with racial or ethnic superiority or inferiority. On the micro-scale, human development is a predictable process that moves through the principal stages of infancy, childhood, adolescence, and adulthood. Just so, on the macro-scale. Root Races refer to landmark developmental stages in the consciousness of humanity each Race bringing forth latent aspects of consciousness over the long course of evolution.

As said above, present-day mankind is cycling forward in order to cycle back on a higher turn of the spiral. Having lost in spirituality that which mankind acquired in physical development until almost the end of the Fourth Root-Race, mankind, now in the Fifth Root-Race, is gradually losing physicality in order to regain once more his former spiritual faculties. This process must go on until the period which will bring in the Sixth Root-Race.

The Bulgarian Master Peter Deunov, spoke of the transitional state between two conditions. He said in 1944: 'We are crossing the boundary between two epochs [...] and entering the new epoch.' The 'new epoch' refers to the transition from Kali Yuga, the fourth and present age of the world cycle of yugas or 'ages'. It is the end of the four ages that comprise a cycle and is often referred to as the 'dark age'. The Kali Yuga leads to eventual destruction of the world order before the creation of a new cycle of the four yugas.

Deunov continues: 'People all over the world, of all nations and races, are now forming the nucleus of a new race, with a new understanding. In the future there will be many more people of the sixth race on the Earth than there are at present. The spiritually wise people throughout the world will form the sixth race. The people of the sixth race will have correct features. They will be beautiful; their beauty having been formed by the high ideal they hold within themselves. They will be far more beautiful than the people of today [i.e., the fifth race].'[16]

However, the new humanity entailing fellowship between all nations will not emerge without painful birth pangs. This has happened in the past like the transitions between the Third and Fourth and the Fourth and Fifth Root-Races and is bound to happen during the transition from the Fifth to the Sixth Race in the form of large scale geological and environmental disasters entailing immense suffering in order to awaken and purify human consciousness, as Blavatsky and Deunov both forewarn.

Stories of miracles performed in ancient times always run the risk of being belittled by sceptics. But it is not so easy to dismiss in our prosaic, modern days authentic reports of individuals with powers and perceptions so extraordinary that no amount of scientific theorising can ever account for them, explain them away, or disprove their existence.

In India we have the example of the legendary sage Ramana Maharshi (1879–1950) who could materialise his presence to his disciples anywhere in the world and then just as soon as the task was accomplished, vanish into 'thin air'.[17] As previously stated, Peter Deunov was able to appear in different places and then disappear without a trace. On other occasions he was able to sense, remotely, a situation of extreme danger and materialise his presence to save the situation.

Hallmarks of a Genuine Spiritual Experience

The overriding question is whether or not a spiritual experience has radically transformed a person for the better—towards less self-centredness and a greater degree of universality of outlook. There are several distinguishing characteristics. First, an experience that touches our innermost centre, however fleetingly, leaves an afterglow, the memory of which is never forgotten and can be summoned up as a feeling in the heart to sustain one during moments of crisis.

Crucially, a spiritual experience radically transforms the entire outlook and conduct of a person, who subsequently may lose all interest in financial acquisitiveness and devote his energies and his wealth towards philanthropic or charitable projects undreamt of beforehand. Quite often, such a person also loses all fear of death, having received intimations of his immortality along with an overwhelming sense of unity.

Anyone who has been blessed with a genuinely spiritual experience is no more likely to talk about it in public than he would discuss the intimate details of his own private and personal relationships. A spiritual experience is a most sacred affair and what is sacred is silent and private other than perhaps to one's closest confidant.

Techniques of Ascent

There are specific and entirely safe meditation practices for attaining the lofty states described above—by slow degrees. These involve meditations on the mystery of the breath, the heart and the eye, all underpinned by unremitting self-enquiry, mental discipline and the cultivation of the utmost refined feelings.[18]

One fine example of the flowering of latent powers in man, gifted to rare individuals, is Helena Petrovna Blavatsky. She also had extraordinarily developed occult powers which, it would seem, enabled her to access the 'astral light', or universal memory, which records everything that has ever happened during this cycle of the history of the world. Its cosmic counterpart, the everlasting Universal Memory, is generally referred to as the 'Akashic Records'.

It was by drawing on this astral light memory that her adept instructors were able to show her the copious references she needed for her works, such as *Isis Unveiled* and *The Secret Doctrine,* which she wrote down during long and exhausting sessions often involving periods of great mental exertion. There is no other way to explain how a woman generally in serious ill health, working in solitude, with a small personal library, could quote extensively from literally several thousands of volumes on every department of human knowledge ranging from archaic manuscripts to modern publications.

Why are such phenomenal powers gifted only to rare individuals? The chief reason is that knowledge is power; and the greater the knowledge, the greater the power; and the greater the power, the greater its benefit for the service of humanity or its abuse for the aggrandisement of the ego in its craving for selfish power, control, and sensual gratification at the expense of the untold suffering of humanity. That there are but few people able to use immense power wisely is an understatement (look at the behaviour of most powerful politicians, billionaires, and dictators).

As Blavatsky potently reminds us: 'In occultism, a most solemn vow has to be taken never to use any powers acquired or conferred for the benefit of one's own personal self, for to do so would be to set foot on the steep and treacherous slope that ends in the abyss of Black Magic. I have taken that vow [...]. I would rather suffer any tortures than be untrue to my pledge.'[19]

Man's Limitless Evolution

Just like those of extraordinary genius each one of us shall also have to successfully negotiate our own individual and chosen path. In fact, there comes a time when we realise that our evolutionary journey is not an option at our bidding but mandatory. Why? Because there is that essential and unalloyed element within each person that seeks to grow from a bud into a blossom. However, an unfinished journey needs to start or continue in the right direction. This means a clear understanding of what is meant by evolution.

But the problem we face nowadays is that, other than with cults like fundamentalist creationists, evolution has become virtually universally accepted currency with the general public, goaded by a simplistic and one-sided rendering by science on the grounds of supposedly proven facts, that 'man came from the apes'. Evolution has therefore become effectively a platitude and a 'catch all' word, its true meaning now blurred by both overuse and misuse.

With a few exceptions, neo-Darwinism is virtually the sole dictum of the scientific establishment on the theory of evolution. That descent occurred through genetic mutation and natural selection and man descended from a hairy quadruped.

Evolution is naturally an enormous subject with which science has been grappling for over one and a half centuries since Darwin published *On the Origin of Species* which, at the time, was a long overdue scientific acid to corrode the accumulated rust of theological dogma regarding the genesis of man. Nonetheless, such scientific endeavour is limited strictly to the changing physical forms. Science and Darwinism look purely to the change in outer form to the virtual exclusion of the innate potentiality of the entity—its purpose, plan, and design.

The occult view of evolution can be summarised in the Qabbalistic axiom: 'A stone becomes a plant, a plant an animal, an animal a man, a man a spirit, and a spirit a god.' Or in Occult catechism quoted earlier: 'I slept in the stone; I stirred in the plant; I dreamt in the animal; and I awoke in the man.'

The real principle of evolution—as occult science affirms—is that *all* life (not merely biological life) is in various stages of *becoming*. There is an evolution of *kingdoms of nature* from the mineral kingdom to the human kingdom, and far beyond. Hence, the 'I', referred to in the Occult Catechism above, is none other than the Universal and Divine Consciousness that seeks ever newer and more complex vehicles and forms—kingdoms of nature—in order to express in ever greater degrees its limitless potentiality. The death of the form therefore does not mean the extinguishing of consciousness. On the contrary, it means that consciousness 'sheds' outworn forms that no longer serve its ongoing evolutionary march.

Progressive modern science is slowly corroborating a fundamental tenet of occult science. This is its discovery of the ascending grades of embodied consciousness in the ladder of life. This means that at any level in the universe, from the lowest to the highest, a parent body—whether that be minerals or primitive life forms or the cells in plants, animals and humans—is host to lesser lives that are existing and evolving through it.

Whether we consider kingdoms of life that are subhuman, human or superhuman, each unit of life lives within the field of the greater life, its host, evolving towards the next stage. The human being, at all levels, is host to countless lesser lives all evolving towards the next higher stage in the evolutionary ladder. Just so, human beings live within, and evolve towards, the overriding solar consciousness and so on.

What stands out is that *human evolution cannot be split off from the evolution of other kingdoms of nature*. It is part and parcel of the total process. Man, like all other evolving entities, has everything in him that the cosmos has because he is an inseparable part of it, its child—the microcosm of the macrocosm. Everything that is in the cosmos is in him, active or latent, and evolution is the awakening and bringing out of what lies dormant within. It is a truism that any evolving entity can become only what it is in its essential nature.

So, from the standpoint of the Perennial Philosophy, the core meaning of evolution is that the nucleus of every entity is a spirit (technically, a monad) that expresses its powers and faculties, through the ages, in matter. The various bodies which change and improve in refinement and complexity as the ages pass—enabling spirit to progressively express a greater proportion of its innate potentiality on any plane of manifestation. For man, these vehicles comprise not just the physical body but the subtle bodies of consciousness.

At the lowest point of cyclical descent, when the greatest materiality is attained, a return cycle is commenced towards a greater expression of the spirit-side as a progressively greater development of spirituality is being achieved. This is the evolution of spirit occurring concurrently with a progressive recession of materiality: the involution of matter. The terms 'evolution' and 'involution' thus apply to *both* spirit and matter on opposite sides of the cycle.

The Antiquity of Man – the Evidence from Archæology

This subject cannot be divorced from that of lost continents of immense pre-historical antiquity on which the early races of man flourished. Whereas most geologists deny the existence of ancient continents such as Gondwanaland, Lemuria, and Atlantis, or relegate them to the realm of folklore, their existence is not without evidence from geology— and in abundance from the ancient records of the Mystery Teachings.

Regarding Gondwanaland, *'Long-Lost Continent Found Submerged Deep Under Indian Ocean'* was the heading of an article in the 31 January 2017 edition of *New Scientist*. Traces of the continent were found in the volcanic island of Mauritius, five hundred miles east of Madagascar. Whereas the location, duration and formative forces behind lost continents continue to be matters of lively scientific debate, the fact of their existence is becoming increasingly problematic to deny. Our purpose, however, is not to delve into this fascinating subject but to point to scientific evidence for the enormous antiquity of the early human races that inhabited these prehistoric continents.

The definitive modern scientific evidence comes from Michael Cremo (*b*.1948), member of the World Archæological Congress, international university lecturer, and invited speaker at many prestigious scientific institutions, such as the Anthropology Department of the Russian Academy of Sciences in Moscow, the Archæology Department of the Ukrainian Academy of Sciences, and the Royal Institution in London. His books *Forbidden Archeology*[20] and its condensed edition, *The Hidden History of the Human Race,*[21] co-authored with the American mathematician and evolutionary biologist Richard L. Thompson (1947–2008), contain a massive number of detailed facts and figures assembled with meticulous scholarship.

Cremo and Thompson have found that over the past 150 years, archæologists have discovered huge amounts of evidence in the form of human footprints, human skeletal remains and human artefacts that are tens, or even hundreds, of millions of years old, going all the way back to well over 2 billion years. In contrast, current textbooks on archæology, based on the Darwinian paradigm that modern humans appeared between 100,000 and 150,000 years ago, mention no such findings or evidence. Nevertheless, the latest pronouncement from mainstream science pushes the origin of anatomically modern humans further back.

It appears that the oldest human artefacts can be dated back to some 2.8 billion years. They are metallic spheres, one or two inches in diameter, recovered a few decades ago from a mine in South Africa near Ottosdal in the West Transvaal region. Metallurgical analysis shows them to be made of hematite, a naturally occurring iron ore. However, what is certainly not naturally occurring are the parallel grooves around the centres of these spheres. Some spheres have four grooves, others three, two and some only one. The metallurgists who examined them said they were not produced naturally. Therefore, the objects must have been manufactured by someone with humanlike intelligence. Yet they were found solidly embedded in mineral deposits over 2.5 billion years old.

Another discovery demonstrating human antiquity going back to over 500 million years is the shoe print found by an American draftsman and amateur fossil collector William J. Meister Sr (1904–1987) in 1968 during an expedition near Antelope Springs, Utah.[21] The print appears to have been made by a human wearing some type of primitive shoe or sandal 10¼ inches long and 3½ inches wide.[22] When Meister split open a two foot slab of rock, the footprint impression was revealed within the slab. It appears that the human that left the imprint stepped on a living trilobite (now a fossil group of extinct marine arthropods).

Meister claimed that when he had the print examined by a geologist, he was offered $250,000 for it. But when Meister asked him, 'What are you going to do with it if I sell it to you?' the geologist replied, 'I'm going to destroy it; it destroys my entire life work as a geologist.'[23]

In 1979, footprints were found in Tanzania by the British palæoanthropologist Mary Leakey (1913–1996). Known as the Laetoli footprints, they were found in layers of solid volcanic ash and dated to 3.6 to 3.8 million years by the potassium-argon method.[24]

The Italian geologist Giuseppe Ragazzoni (1824–1898) discovered anatomically modern human skeletal remains of a woman, a man and two children from a Middle Pliocene formation in Castenedolo in northern Italy—and the Pliocene dates from about 2 million to 5 million years ago.[25]

It bears repeating that teachings on the antiquity of the human race, its origins and devolution are at the core of all esoteric and spiritual instruction. They are to be found in the Qabbalah, Gnostic Christianity and the doctrines of the ancient Egyptians, American Indians, and Scandinavians. However, it finds its most forceful, clear and detailed account in the Vedas. Written in Sanskrit, they mention a human presence that goes back over 2 billion years on Earth.

The terrestrial human body is a vehicle to awaken the conscious self to an awareness of its original state in a realm of pure consciousness where spiritual human archetypes have always existed. Human devolution is the process by which conscious selves enter, or take on, a material form, i.e., in human bodies on Earth. This process has been going on for vast periods of cyclic time not only in this universe but in countless other universes where consciousness expressed through the human kingdom has been manifest for aeons but not necessarily in the human *form* as on Earth. If this sounds too far-fetched it is only because of our unduly anthropocentric outlook, inculcated by modern science, which seems to regard the human species as one of those inexplicable accidents that happened to occur on a supposedly 'minor planet of a very average star'.

The evidence for the antiquity of humanity on Earth, and that the human state exists in other domains of the kosmos, can all be found in sources like: the archaic stanzas of occultism and in the sacred literature of Hinduism. As Cremo explains, 'the Vedic writings speak of some 400,000 humanlike species scattered throughout the universe.' Moreover, he continues, 'anatomically modern humans and the various hominids, such as the Australopithecines, could be placed among those 400,000 species.'

The *Ramayana* mentions the *Vanara*, a species of ape-like men that existed millions of years ago.[26] *But alongside these ape-like men existed humans of our type. The relationship was one of coexistence rather than evolution*. It is stressed that these ape-like men bear no relation to the simian or anthropoid primates from which, according to Darwinism, present humanity is supposed to have evolved.

It is illogical even to suggest that a higher entity—man—can evolve from a lower entity, a hairy quadruped—by *descent*. Descent from such a lowly creature can only result in even cruder versions of the same type, that is, a lower entity not an ascent to a higher entity—man.

The Question of Intelligent Design

The universe should show signs that it was designed by higher intelligences for accommodating humans and other forms of life.

To avoid the uncomfortable question of intelligent design and a designer some scientists resort to blind chance. The futile example often cited is that a monkey seated at a keyboard could, by pure chance, faultlessly reproduce a Shakespeare sonnet. The chance of a monkey typing merely the word 'Monkey' is calculated to be 1 in 1.5×10^{17}—a figure which is the same as the approximate age of the universe in seconds.[27] In fact such behaviour contravenes the Second Law of Thermodynamics, which paraphrased is that monkeys will always behave like monkeys, leaving lofty literature to Shakespeare.

So there is no satisfactory explanation—other than blind chance—of how an entire universe (including human beings whose characteristics range from unspeakable cruelty to ineffable genius) manifesting according to pure chance could, in its ongoing evolution, display such beauty, dynamism, structure, and organization.

Here we experience a head-on collision with creationists and fundamentalists who maintain that since chance cannot account for the fine-tuning of the universe and all life in it, this shows the handiwork of intelligent design by God, the Intelligent Designer. It is no surprise that the most vociferous amongst the intelligent design fundamentalists are Christian. But design by a designer begs the question of who designed the designer and so on.

Like the Darwinists, the intelligent design advocates accept that humans and other living entities are merely complex forms of matter. The only difference between the Darwinists and the intelligent design theorists is how the complexity came to be. The Darwinists attribute the complexity to evolution by genetic mutation and natural selection whereas the intelligent design proponents attribute it to intelligent design by an extrinsic 'Creator God' who, presumably, somehow managed to design Himself by Himself.

Many cosmologists admit that the odds against the fine tuning are too extreme for simple chance to be offered as a credible scientific explanation. But to avoid the question of a designer they have proposed the multiverse—a practically unlimited number of universes mentioned previously. In each the value of the fundamental constants and laws of nature are adjusted in a different way—and we just happen to live in the one universe with everything adjusted correctly for the existence of human life. Indeed, the Australian physicist Neil Manson (*b.*1940) of the Australian National University has said, 'The multiverse hypothesis is alleged to be the last resort for the desperate atheist.'[28]

Occult cosmology (including the Vedic) also speaks of 'numberless Universes' but all of them are designed for life *at their level* and through the appropriate substance for the expression of life at that level. But beyond all of these material universes or globes, with their levels of gross and subtle matter, is the level of pure consciousness or spirit.

A Child of Cosmos

A high-grade substance produced from something crude cannot happen by itself. There has to be a refining principle at work. In the case of man, it is the Higher Self seeking ever more refined and complex bodies and personalities through which to manifest across innumerable incarnations. So, if barbarians are to evolve towards cultured people, this principle has to operate.

Natural selection and genetic mutation are supposed to ensure the survival of the fittest. But that could apply to ensuring the survival of just the fittest amongst the barbarians. Why do they evolve towards a higher order of life and complexity? What is the ever-active refining principle in man? It is the evolution of consciousness. Man has descended from Divine Consciousness (Spirit). So also has cosmos. Therefore, man has in him everything that cosmos has because he is an inseparable part of it.

Man is the child of cosmos and man cannot be separated from his parent. Every principle that is in cosmos is in him, latent or active, and evolution is the bringing forth of what is within. As periodically stressed, the growth and refinement of man is quite literally, limitless. Therefore, man cannot possibly be 'descended from some lowly-organised form', or 'descended from barbarians.' Nor indeed can the unfoldment of consciousness in man be merely a passive process. It also requires aspiration and directed effort.

Occult science complements and corrects errors in the scientific picture by virtue of taking in the wide sweep of evolution on all kingdoms of nature, *especially the inner worlds and their causative influence on the outer*. Regarding man, this includes evolution on the mental and psychic planes in addition to the physical. In contrast, biology and evolutionary theory are concerned entirely with the outer forms constituted of physical atoms and molecules and the physical and electro-chemical forces acting between them. Only a few biologists will uneasily and privately admit the notion of vitality which is present in all living forms.

By revealing the missing link and supplying the missing gaps in knowledge, occult science enables a much richer and deeper understanding of evolution and elucidates the purpose and planned progression of the whole evolutionary process without which the evolution of forms would be a futile, chance endeavour by nature.

Reincarnation is a key occult doctrine and central to the evolution of man. However, we face two major hurdles. First, the doctrine of reincarnation is so completely misunderstood, especially in the West. Secondly, the role, function, and imperative of reincarnation cannot be appreciated in its fullness without invoking other doctrines that are not synonymous, but to which it is strongly related. These are principally, the *Doctrine of Essential Identity and Rebirth,* or the *Doctrine of Constant Renewal,* aspects of which are the doctrines of *Transmigration* and of *Metempsychosis*. These, incidentally, have nothing whatsoever to do with a human being born again as an animal (or stone, etc.) in the next life, which is a grotesque distortion of sublime doctrines.

The occultist is one who is able to perceive truths concealed in the world religions, sciences and philosophies from this higher dimensional standpoint. Similarly, the individual, great (but not necessarily famous) scientist, often working alone, comes to his ground-breaking discoveries by virtue of not being hidebound by orthodox concepts and, therefore, better able to see further into the nature of things than establishment scientists.

The understanding that comes from a higher dimensional standpoint is organic and holistic instead of linear and sequential. Therefore, the occultist can perceive relations and underlying causes before their appearance in the ordinary space-time world of four dimensions. The phenomena of clairvoyance and remote viewing have all to do with such altered states of consciousness.

It is the writer's strong conviction that the explanation of paranormal phenomena, psi, apparitions and a host of psychic phenomena, lie in higher dimensions above the familiar four-dimensional world of space-time we normally inhabit. Here, physics has much to offer in terms of superstring theory and M-theory that invoke higher dimensions An extremely comprehensive, albeit technical exposition on this subject is in *The Mathematical Connection Between Religion and Science*.[29]

The Big Questions

There can be few people at some stage in their lives, invariably in moments of distress or crisis, boredom or world-weariness, who do not ask questions like: '*What makes my life worth living? What happens after I die? Why am I in such pain? What is the meaning behind all the cruelty and sorrow in the world? What place has love and joy in a world full of evil and injustice? What goals should I choose? Do I have a place in the overall scheme of things, if indeed there be such a scheme? How can I discern what is true from what is false? What is knowledge? Is everything I experience just a matter of luck, chance or coincidence? If there is supposed to be a loving God, then why am I unloved and suffering in loneliness?*'

The above list of doleful self-questioning could be extended indefinitely but all questions are part and parcel of one big question, and they all coalesce towards a single, central problem: that there are times when a person feels either pinned down to a hard rock or like a feather blown around in the wind. We need not labour the point that the better we know ourselves—and thereby, our fellow beings and the world we live in—the more will our lives have purposeful energy and direction during adverse circumstances and the more intelligently will we navigate the reefs and rocks of doubts and difficulties.

For that reason, the more important but less common question that few bother to ask of themselves is: '*Who* is asking the question?' Pondering over this issue may reveal the difference between who we truly are and who we think we are; in other words, the difference between our Self and our self-image. And thereafter to the realisation that the answers lie within each of us. Nonetheless, we have constantly emphasised that such self-knowledge must be multi-levelled and involve all planes on which man lives and has his being. It is evident that a knowledge of physical man, obtained through science, is as indispensable to the understanding of man as the insights into his soul and spiritual nature revealed by the Perennial

Philosophy. That is why the epigraph to this chapter declared, 'between the shores of the ego and the ocean of the soul there are many planes of consciousness.'

The Janus Faces of Science

In ancient Rome Janus was the god of beginnings, endings, and transitions. Janus is depicted with two heads, one looking forward into the future and the other looking backwards into the past. In order for science to unfold its full potential it needs to look to both the future and the past, drawing upon all that has gone before in order to build a new and better informed world.

Although, it is undoubtedly true that mainstream science has become tainted by a nihilistic paradigm wholly foreign to its true nature, we need to peel back those accretions to reveal true science, simultaneously learning from our past mistakes. It is notable that Janus in himself combines both future and past and, in a sense, transcends both in the ever present now.

Without this understanding of the 'dual embrace' in the unity of the now, the Janus face turned to the past becomes enmeshed in outmoded thoughts and actions, a fearful submission to establishment authority and an inability to be receptive to new evidence. The face turned to the future becomes equally handicapped by failing to profit from the wisdom of the past resulting in past errors being repeated due to the same mindset that led to them.

Quantum physics has broken down not only the entrenched concepts of classical, materialistic physics but also its philosophical implications: that an intrinsic unity exists at a deep level of the universe, that what we call 'matter' is but a conglomeration of forces and the interconnection of all parts of the universe. This has been shown by rigorous theory and punctilious, repeated experiments. Whereas the philosophical implications of the rarefied ideas of the latest physics have still to pervade the collective consciousness of mainstream science (especially in biology), it can only be a matter of time before this happens.

We can then be in no doubt that when the paradigm adopted by mainstream science changes from materialism toward holism, from objects and matter towards consciousness and mind, the media will (hopefully) swiftly report this radical change in outlook and the public understanding of science will be much elevated.

This is already happening, albeit slowly. By its own methods, research, and discoveries, avant-garde science is now slowly transcending the materialist philosophy, presaging the threshold of a major spiritual revival and pointing towards new meaning in a living world. Scientific studies of a range of spiritual practices reveal that they have remarkable benefits for health of mind and body and give a greater sense of connection with a transcendent reality which goes above and beyond the confines of the ordinary human personality. Such practices include meditation and prayer, service and gratitude, rituals and pilgrimage—all now widely available to people of every religion and to those who uphold no orthodox religion.

This theme of spiritual practices relevant to a scientific age is in *Science and Spiritual Practices*[30] and *Ways to Go Beyond,*[31] two of the latest books by Rupert Sheldrake, an all- too-rare individual who seamlessly integrates cutting edge science with spiritual sensitivity and has the courage to promulgate his ideas dispassionately in stark contrast to the rhetoric and intemperate attacks upon him by his detractors.

Science has also widened and elevated the consciousness of humanity by forcing a shift from the 'village mentality' to a global outlook. This has come through such factors as international trading agreements, globalisation of world markets, and high-speed travel. Near instant communication between people, news reporting and the ubiquitous internet has 'shrunk' the world onto our computer screens. Such incredible facilities are the products of advanced technology—the fruit of scientific materialism! Obviously, these science-driven opportunities have created massive problems in their wake—like damage to the ecosystem and electronic hacking—which science itself will have to sort out. Nonetheless, it is suggested that the recognition of global brotherhood which currently exists, primarily at the economic and material level, is an unstoppable trend which will gradually flower, in the fullness of time, towards an understanding and realisation of the spiritual affinity amongst mankind and nations and between man and nature.

In essence, science has not only enlarged the sphere of its own disciplines but assimilated the experiences of mystics.

This sea change in modern science has been possible because of the tremendous advances in neuroscience and brain mapping techniques that have clearly demonstrated the changes that occur in both brain state and brain structure during profound states of meditation. Indeed, Oneness, the deep interrelationship of life and the dynamic nature of reality are factors that have *always* been the experience of advanced mystics, but they are now being confirmed by science, especially through the life sciences and ecology. So, for all the above reasons, science has earned its respect having gained the upper hand over other ways of knowing. It has rightfully earned its status by the great value and utility of its practical applications, as well as its methods, leading to results that are generally factual, repeatable, and accurate.

The Dark Face of Science

'Are we just useful parts of a purposeless machine operating in the midst of randomness?'[32] asks Vasileios Basios (*b.*1962), the Greek physicist specialising in chaos theory and the physics of complex systems. Indeed, we are, according to many internationally famous scientists. It appears that science has created its own prison—but science has also created the key to unlock it. So, it boils down to science appreciating its legitimate playing field which necessarily means working within its boundaries and context. In this wise, science reigns supreme in its understanding of physical mechanisms and behaviour, whether of the body of man or the visible and objective universe. But any attempt to understand the destiny of man or the purpose of cosmos will invariably prove unsatisfactory without a concomitant understanding of the nature of mind.

How do celebrity scientists view religion and philosophy? Nobel physicist Steven Weinberg FRS (1933–2021) claims that, 'Religion is an insult to human dignity' justified on the grounds that, 'With or without it you would have good people doing good things and evil people doing evil things. But for good people to do evil things, that takes religion.'[33] And he adds that apart from a few notable exceptions, 'Most physicists today are not sufficiently interested in religion even to qualify as practicing atheists.'[34]

Given that Weinberg has ignored the best ideas in philosophy and religion, arbitrarily confined his enquiry entirely to materialistic concepts which *do not represent the best ideas in modern physics*, and has disregarded the metaphysical dimension of that subject for which he was awarded the Nobel Prize—quantum physics, which points to the primacy of consciousness—his message of bleak despair should hardly surprise us. It is a perfect example of the airless cul-de-sac of materialistic science grappling with concepts completely beyond its ken.

The New Black Magic of Science

As Manly Hall so astutely points out, scientists conveniently forget that although the demonism of the Middle Ages, which they take such delight in scorning, seems to have disappeared, there is abundant evidence that the black arts, or '*black magic has merely passed through a metamorphosis, and although its name be changed its nature remains the same* [emphasis added].' Witness the many forms of so-called 'prosperity thought', 'will-power building', high-pressure salesmanship.[35] (Bearing in mind that Hall penned his remarks in the 1920s; one shudders to think of what he would have said about much of the contemporary scene in science.) Add to that, the abuse of nuclear energy and the ghastly experiments on animals in the guise of medical research.

Add to these abnormal genetic engineering like producing a mouse from two female parents,[36] transplanting organs or tissues between members of different species,[37] a human head transplant on a corpse,[38] the transhumanism and cryonics movements, the radicalisation of the internet by religious fanatics and the repellent list of black arts, using modern scientific technology *unethically,* seems depressingly endless.

Modern Man's Spiritual Crisis

Modern science has bequeathed humanity with precious gems—for which humanity has had to pay a terrible price. On the one hand, as we have outlined, science has been instrumental in alleviating untold poverty and physical suffering and has given us living conditions, medicines, hygiene, and transport undreamt of a few centuries ago. Anyone who visits his dentist for a tooth extraction—let alone undergoes major surgery—should fall on his knees in gratitude for the discovery and skillful use of anaesthetics. So also do we owe an enormous debt of gratitude to neuroscience and brain surgery.

But on the other hand, the price extracted has been, arguably, even worse than the legacy of modern weapons of mass destruction, man-made ecological disasters or the assault upon nature and the countryside. And that price is the *spiritual* crisis of modern man, the suffering and poverty of his soul, his impersonalised and dehumanised existence aided and abetted by the alarming increase in artificial intelligence devices. There is his feeling of social isolation and dislocation, his feeling that he is merely a utilitarian cog in a giant economic power machine and the treadmill and drudgery of his daily existence. All this, and more, through the dead hand of entrenched materialism feeding him with the numbing idea that we human beings are *merely* animals, and that both humans and animals are, in the final analysis, *just* complex biological machines and nothing else.

The English philosopher Nicholas Maxwell (1937–2024) has devoted much of his working life to arguing that there is an urgent need to bring about a revolution in academia so as to promote wisdom and not just acquire and accumulate knowledge.[39] This revolution in education is also a principal theme of *Harmony: A New Way of Looking at Our World* by King Charles III.[40] The whole issue is compounded by problems caused by the excessive use and misuse of the internet—short attention span, hyper-consumerism and spectacularly shallow or crude entertainment.

According to the *National Review (New York)*, some 27% of Americans live alone nowadays, compared to 13% in 1960. Moreover, a 2010 survey showed that a third of adults were chronically lonely compared to 20% in 2000. The same article states that white Americans are dying not just of heart disease or cancer, but of diseases that imply a '*sickness of spirit as much as of body—suicide, drug overdoses, and cirrhosis of the liver* [emphasis added].'[41]

In Britain, the statistics are equally depressing. This is what the contemporary British award-winning physician and consultant in integrated medicine Dr Kim Jobst (*b.*1950) calls 'diseases of meaning', in other words, that disease ought to be seen as a meaningful state that can inform health workers how to help patients to heal themselves. In this way, instead of being regarded as meaningless, individual problems become *diseases of meaning*, enabling people to move away from a victim mentality towards a state of empowerment by helping them become stronger, to live more fully and with more understanding.[42]

The plight of young people is now a matter of increasing concern. A recent survey (2018) found that a quarter of girls and one in ten boys were self-harming. Furthermore, a report in the British Medical Journal found that self-harm among adolescent girls in the UK has risen disproportionately by 68% in just over three years and that those young people who self-harmed were at significantly increased risk of committing suicide.[43] *Yet these distressed teenagers are not exactly poverty-stricken and seem fully capable of indulging their every technological and electronic whim via their smartphones, iPads, and such like.* But under the strain of coping in the excessively pressurised times now prevailing (driven largely by social media), life for many of them has become meaningless. Science policy-makers need to act to counteract the increasing dehumanisation and spiritual starvation in society through excessive materialism.

Materialism at the expense of spirituality has been the cause of the majority of these problems. However, materialistic solutions are sought in order to resolve them when, in fact, they should be employed the least.

Hence, there is an inevitable tension caused by the two opposing Janus Faces of science: the one gazing ever onwards and upwards seeking a loftier standpoint and the other staring backwards and downwards steeped in materialism. This tension is not caused by errors in methodology but is inherent within their contrasting worldviews.

It has been assumed that cold intellect represents the zenith of human mental capacity, whereas spirituality has been allowed to fall and wither by the wayside. Accordingly, 'intellectualism' has been artificially elevated to a position of ultimate authority and is supposed to be able to deal with problems that lie completely outside its legitimate boundaries and limited influence.

Again, we stress that in contrast to materialistic science, occultism throws light on the inner driving and guiding forces and intelligence which underlie all natural processes and phenomena and thus makes our conception much more meaningful and richer, albeit sometimes lacking the detailed and technical information forming part of scientific knowledge. Therefore, establishment scientists may strain every nerve and muscle and invest vast sums of research money in their efforts to wrest the secrets of consciousness from nature. But they will never achieve their intended goal until they learn that nature will never divulge her innermost secrets to those who bang loudly at the door that bars the sacred arcana of truth from the profane.

We affirm that the secrets of life, mind, and consciousness will become an open book only to those who qualify themselves in terms of joyful self-sacrifice, deep humility, a spotless morality, reverence for nature and a philanthropic outlook. And all that allied to a profound study of the perennial wisdom balanced with experiments, not in a laboratory with hapless, suffering creatures, but in the 'laboratory of the mind', by reading the 'Book of his own Life'. With rare exceptions, this is undeniably a tall order for most people. However, those who measure up to it are the sages and adepts and they are, indeed, a rare breed since they epitomise the finest attributes of humanity.

Nonetheless, there is no law that forbids anyone from sincerely aspiring towards such an ideal. Therefore, we say that those scientists, still much in the minority, who strive very sensibly to enlarge the domain of physical science by trespassing on the forbidden grounds of mysticism and metaphysics are the pioneers of our generation because, having studied physical nature in depth, these enlightened scientists have sensed that their wonderful discoveries will remain largely misused and ungoverned unless they cross into the domain of occult science and lift the veil of physical matter to see beyond and search for its noumenon.[44]

The appearance of science on the world stage in the present epoch signifies a momentous stage in the evolutionary cycle of nature. The great scientists are the chosen instruments for the discovery or revelation of a few of nature's secrets. Legendary scientists like Newton and Einstein, Schrödinger (1887–1961) and Bohm (1917–1992) have sensed their role as instruments in higher hands and therefore shown a reverence and awe towards nature. A similar dose of humility would not go amiss amongst some of the most famous contemporary mainstream scientists.

Have those scientists who dismiss religion out of hand devoted the same time, energy and effort towards its study over decades in the same way that they have given to science? Understandably, since science is their vocation, they would not have had an equal amount of time to look in-depth at religion, but should they not attempt to gain some understanding of the bigger picture rather than dismissing religion outright?

Does it make sense for doctors, who dedicate several years of study and training in medical schools, to dismiss near-death and out-of-body experiences without sacrificing considerable time studying the evidence? Then take water divining, an age-old practice used by water companies to this day. Unable to explain dowsing on the basis of existing physical science, the practice and the practitioners are jeered on the basis that dowsing is not scientific and so must be a form of witchcraft.[45, 46]

It is a common fatal error, and totally unscientific, to presume that knowledge and authority in one field can automatically be grafted onto and used to evaluate the worth of another totally different field. We find an obvious example of this in medical science where doctors trained exclusively in conventional (i.e., allopathic) medicine will denigrate outright therapies, like acupuncture and homœopathy, on the basis of their concepts about mainstream medicine, without bothering to understand the different standpoints and methodologies of these complementary therapies which—unlike allopathic treatments—are essentially holistic and not reductionist.

Occult science has been referred to as the *Royal Science* because it encompasses all dimensions of reality. Occult science displays the finest attributes of SCIENCE in that every theory or proposition is tested and checked for evidence, obviously not in a physical laboratory, but in the realm of consciousness. Whereas physical science uses microscopes and telescopes as extensions of the physical senses, occult science uses the instruments of unfolded inner faculties of consciousness to transcend the senses.

The challenge to the serious occult student is twofold: first, to acquire a clear idea about the fundamental doctrines in general terms and, secondly, to comprehend their relation to one another as an integrated whole. This is a far cry from the popular conception of knowledge as a linear stockpiling of fact upon fact by accumulating a great weight of data. Ironically, it is the very simplicity of the fundamental occult doctrines that cause the difficulty! And that is because our minds nowadays are so trammelled by detail and complexity that we have difficulty in assimilating doctrines of simplicity that are in direct proportion to their depth and profundity. Simplicity is not always simple or easy to comprehend!

It is only by adopting a holistic approach, by amassing an abundance of proofs from diverse sources all tending to show that in every age, under every condition of civilization and knowledge, the learned classes of every nation made themselves the more or less faithful echoes of one identical system and its fundamental traditions—that readers can be made to see that so many streams of the same water must have had a common source from which they started.[47]

Consciousness Is an Element

It is crucial to acknowledge the primacy of consciousness. Because every idea that we may have, all our dreams and aspirations, joys, and sorrows, in fact, all that we know is experienced by, in and through consciousness. It is only from this starting point that we can make sense of the riddle of life and resolve the problem of subjective and objective reality.

Having established—*through our own experience*—that we humans are conscious, let us move first down the scale, initially to animals and then to plants. Are they conscious? There is every evidence that they respond to human feelings and thoughts, but would that constitute consciousness? If so, then what about birds, reptiles, insects, jellyfish, molluscs, and amoebas? Are they conscious? If not, then where in the vast spectrum of life does consciousness begin and end? Which of the following two arguments seems the more reasonable:

1. That there is some arbitrary line or point of origin that demarcates the starting point of consciousness.
2. Or that there is a graduated scale of consciousness, such that:
   i. the lower down the scale, the lesser is the degree of awakened consciousness;
   ii. conversely, the higher up the scale, the more complex are the evolved vehicles through which consciousness can express its potentiality.

The vast majority of scientists would, in fact, make the case for the first argument above on the basis that the biological cell constitutes the basis of all life on Earth (and possibly even elsewhere in the universe) and therefore constitutes the dividing line between consciousness and non-consciousness, the animate and the inanimate. If that be the case, we must ask what is so special about a cell? Is the cell therefore conscious? The DNA at the heart of a cell is, after all, only a chemical in the final analysis. Then what is so special about a chemical? After all, a chemical is made up of molecules, atoms, and subatomic particles. It only requires an extension of this line of argument to reason that consciousness is present in the cell, in its chemical constituents, in the molecules, atoms, atomic particles, subatomic particles, and so on, until we find that what we call 'matter' is *mental in essence.*

Moving up the scale, why should the Earth not be conscious, as also the Sun and all the celestial bodies in the solar system? Rupert Sheldrake gave a thought-provoking lecture asking us to look beyond the strict confines of scientific materialism and consider how the consciousness of stellar bodies (such as our Sun) is not only of anthropological or cultural interest but a valid field of enquiry in modern philosophy, psychology, cosmology, and neuroscience.[48] But why stop there? Are not also the Milky Way, the distant galaxies and, indeed, the whole universe, seen and unseen, conscious?

There is no boundary between the animate and the inanimate and there is no arbitrary cut-off point in the scale of life from which we can declare that consciousness operates from here onwards in biological life, but not in so-called inanimate, mineral life. The concept of dead matter is a total oxymoron.

As the Qabbalistic proverb again affirms, there is nothing but life, in various stages of being, and becoming. 'There is no such thing as either "dead" or "blind" matter, as there is no "Blind" or "Unconscious" Law. We men must remember that because *we* do not perceive any signs—which we can recognise—of consciousness, say in stones, we have no right to say that *no consciousness exists there.*'[49] The whole universe is teeming with life in infinite shades of being and becoming at all levels from the spiritual to the physical.

Therefore, we state that CONSCIOUSNESS CAME FIRST or in other words, that CONSCIOUSNESS IS AN ELEMENT. Consciousness is the quintessential principle: it is irreducible; it is not the product of anything else; it is not a plurality. As hydrogen is the most abundant chemical substance in the universe and the lightest element in the periodic table, consciousness may be seen as the ever-present primary ('lightest') Element of all manifestation.

The hierarchy from consciousness to matter can be unfolded in two ways, from the viewpoints of 'spirit' and 'matter'; the former in the vocabulary of the Perennial Philosophy and the latter in the terminology of modern science:

1. *REALITY as PURE CONSCIOUSNESS is the primary ELEMENT ➟ Mind its form ➟ Energy–Matter the appearance, or form, taken by Mind.*
2. *REALITY as PRIMAL INTELLIGENT ENERGY–SUBSTANCE is the basic CONSCIOUSNESS-ELEMENT ➟ Mind its expression as Energy ➟ Matter as Energy imprisoned in form.*

The Dark Janus Face of Spirituality

Earlier, we did not mince our words about the dark 'Janus face' of science, due to its failure to learn from the past and move forward into the future. In the same vein we need to be equally forthright about the delusions and confusion caused by those professing spirituality as this too sinks into the quagmire of materialistic ideologies when severed from its ancient roots.

In the world of science and technology the peer review system along with professional institutions exist to vet the qualifications and views of the scientist. Unfortunately, in the spiritual realm there are no professional or academic qualifications to endorse the spiritual attainment and status of an individual, nor can there be. In the realm of subjectivity and the spirit, the only trustworthy 'peer reviewer' is the guiding hand of the *Spiritual* Ego over the thoughts and actions of the individual concerned.

When that guidance is firm the individual behaves ethically and directs his actions impersonally towards engendering universal goodwill. But when weak, the lower undisciplined personal ego can give rise to a whole spectrum of defects ranging from self-importance to delusion and megalomania. This is why the qualifications for occult instruction are 'qualifications of the spirit'—compassion and altruism, over and above (but never excluding) 'qualifications of the brain'—intellect when allied to the higher nature.

It is a colossal mistake to believe that members or leaders of spiritual or esoteric societies are all 'bringers of light' and spiritually a 'cut above' so-called ordinary folk. New Age and esoteric societies can also be shark-infested with highly self-serving and ruthlessly ambitious people. In many ways they are more dangerous than the career- goaded politicians or the financial heads of industry because their motives are clear whereas those of the former are veiled by a mask of spirituality.

A high intellect, when unbridled by ethical and moral restraints, can all too easily degenerate into *cunning,* which is certainly a step in the direction towards evil. Brutality and sensuality are relatively easy to recognise: not so, cunning, because it invariably wears a 'spiritual mask' perfumed with much glamour.

NOTES

1. Revd Professor Stephen Wright, 'Making (Sacred) Space for Staff Renewal and Transformation', *Journal of the Scientific and Medical Network,* 125 (2018), 21.
2. One of the most comprehensive and readable books is I. K. Taimni, *The Science of Yoga: A commentary on the Yoga-sutras of Patanjali in the light of modern thought* (Adyar, Madras: Theosophical Publishing House, 1965). Taimni was Professor of Chemistry at the University of Allahabad in India, an influential scholar in the fields of Yoga and Indian Philosophy, and a leader of the Theosophical Society.

3. Eben Alexander and Karen Newell, *Living in a Mindful Universe: A neurosurgeon's journey into the heart of consciousness* (US: Rodale, 2017), 38.
4. R. Douglas Fields, 'Brain Reactions', *RSA Journal*, 4 (2016–2017), 36. This article is based on *Why We Snap: Understanding the rage circuits in your brain* (New York: Dutton, 2016).
5. Rephrased from I. K. Taimni, *Man, God and the Universe*, 432–3.
6. *SD*-I, 'Proem', 16–17. Further elucidated in Edi D. Bilimoria, *The Snake and the Rope: Problems in Western science resolved by occult science* (Adyar, Madras: Theosophical Publishing House, 2006), 256.
7. John Maddox, 'The Prevalent Distrust of Science', Nature, 378 (1995), 435.
8. J. D. Watson and F. H. C. Crick, 'A Structure for Deoxyribose Nucleic Acid', *Nature* (25 April 1953), 171, 737–8 [repr. online] <https://www.3dmoleculardesigns.com/3DMD-Files/DNA-Discovery/PDFs/AnnotatedWatsonandCrickpaper.pdf> accessed 28 December 2020.
9. A wisecrack by the writer at the Crick Manifesto—see Francis Crick, *The Astonishing Hypothesis: The scientific search for the soul* (New York: Charles Scribner's Sons, 1994), 3.
10. Dr B. W. Richardson, 'Theory of a Nervous Ether', *Popular Science Review* (repr. London: Forgotten Books, 2019; citation from 1st edn 1871), x, 380–3. Quoted also in *SD*-I, 'Life, Force, or Gravity', 531.
11. Annie Besant, *An Introduction to Yoga* (Adyar, Madras: Theosophical Publishing House, 1972).
12. K. S. Cohen, *The Way of Qigong: The art and science of Chinese energy healing*, foreword by Larry Dossey, MD (USA and Canada: Random House Publishing, 1997).
13. *SD*-I, 'Stanza I: The Night of the Universe', 45.

14. *SD*-I, 'Stanza VII: The Parents of Man on Earth', 224–5.
15. 'Constitution of the Inner Man', in H. P. Blavatsky, *Studies in Occultism* (Point Loma, California, The Aryan Theosophical Press, 1910) <https://www.theosociety.org/pasadena/hpb-sio/sio- cons.htm#t1> accessed 21 April 2020.
16. Beinsa Douno, *The Teacher*: Volume One – *The Dawning Epoch* (London: Shining Word Press, 2016), 119 n.1, 205 n.22, 127 n.11.
17. Once such instance is recounted in Paul Brunton, *The Secret Path: A technique of spiritual self-discovery for the modern world*, foreword by Alice A. Bailey (London: Rider & Co., 22nd impression, 1934), 20–1.
18. Paul Brunton, *The Quest of the Overself* (London: Rider & Co., 1937).
19. Constance Wachtmeister, *Reminiscences of H.P. Blavatsky and 'The Secret Doctrine'* (London: Theosophical Publishing Society, 1893), 46. Quoted also in: Geoffrey A Barborka, *H. P. Blavatsky – the Light-Bringer*, the Blavatsky Lecture 1970 (London: Theosophical Publishing House), 41.
20. Michael A. Cremo and Richard L. Thompson, *Forbidden Archeology: The hidden history of the human race* (Los Angeles, California: Bhaktivedanta Book Publishing, 1993).
21. ——*The Hidden History of the Human Race: The condensed edition of Forbidden Archeology* (Los Angeles, California: Bhaktivedanta Book Publishing, 1999).
22. 'The Meister Print', *Footprints in Stone*, <http://www. footprintsinstone.com/the-footprints/meister- print> accessed 2 April 2020.
23. 'The Meister Print', op. cit.
24. Michael A. Cremo and Richard L. Thompson, *The Hidden History of the Human Race*, 261–4.
25. ——*Forbidden Archeology*, 422–32.
26. *Adhyatma Ramayana: The spiritual version of the Rama Saga* (original Sanskrit with English translation by Swami

Tapasyananda) (Madras: India, Sri Ramakrishna Math, 1985). A highly readable version of the epic is by Channing Arnold, simplified by Marjorie Sykes, *The Story of the Ramayana* (Calcutta, Bombay, Madras: Orient Longmans, 1951).

27. See Edi D. Bilimoria, *The Snake and the Rope: Problems in Western science resolved by occult science* (Adyar, Madras: Theosophical Publishing House, 2006), 210.
28. Neil A. Manson, 'Introduction to God and Design' <http://home.olemiss.edu/~namanson/G&Dintro.pdf> accessed 3 April 2020.
29. Stephen Phillips, *The Mathematical Connection Between Religion and Science* (Eastbourne, UK: Anthony Rowe Publishing, 2009), 45–59.
30. Rupert Sheldrake, *Science and Spiritual Practices: Reconnecting through direct experience* (UK: Coronet, 2018).
31. ——*Ways to Go Beyond and Why They Work: Spiritual practices in a scientific age* (UK: Coronet, 2019).
32. Vasileios Basios, in 'Endorsements for the Galileo Commission Report', *The Scientific and Medical Network.* <https://archive.galileocommission.org/category/people/roland-benedikter> accessed 5 April 2020.
33. Steven Weinberg, Address at the Conference on Cosmic Design, *American Association for the Advancement of Science* (Washington, DC, April 1999).
34. Steven Weinberg, *Dreams of a Final Theory* (London, Hutchinson Radius, 1993), 205.
35. Cited and elaborated from *STA*, 'Ceremonial Magic and Sorcery', CI–CII.
36. Henry Bodkin, 'First Mammal with Two Mothers is Born as Gene-Editing Breakthrough Creates Mouse with No Father', The Telegraph, 11 October 2018 <https://www.telegraph.co.uk/science/2018/10/11/needs-men-first-mammal-born-same-sex- female-parents-gene-editing> accessed 8 April 2020.

37. Jennifer Leman, 'Baboons Survive 6 Months after Getting a Pig Heart Transplant', Science News (5 December 2018) <https://www.sciencenews.org/article/baboons-survive-6-months-after-getting-pig-heart-transplant> accessed 8 April 2020.
38. Sarah Knapton, 'World's First Human Head Transplant a Success, Controversial Scientist Claims', The Telegraph, 17 November 2017 <https://www.telegraph.co.uk/science/2017/11/17/worlds-first-human-head-transplant-successfully-carried> accessed 8 April 2020.
39. Nicholas Maxwell, *How Universities Can Help Create a Wiser World: The urgent need for an academic revolution* (UK: Imprint Academic; USA: Ingram Book Company, 2014.
40. HRH The Prince of Wales, Tony Juniper, and Ian Skelly, Harmony: *A new way of looking at our world* (London: Harper Collins Publishers, 2010). See also Edi Bilimoria, 'The Quest for Harmony: A Unifying Principle in Spirituality, Science, Sustainability and Healthcare', review of conference for the 70th birthday celebration for HRH the Prince of Wales, in *Journal of the Scientific and Medical Network,* 129 (2019), 26–8.
41. National Review, 9 December 2016 <http://www.nationalreview.com/article/442901/white-american-death-rates-rising-life-expectancy-declining> accessed 24 June 2020. See also The Week, 7 January 2017, 14.
42. Kim A. Jobst, Daniel Shostak, and Peter J. Whitehouse, 'Diseases of Meaning, Manifestations of Health, and Metaphor', *The Journal of Alternative and Complementary Medicine,* 5/6 (September 2007).
43. *Guardian,* Editorial, 29 August 2018.
44. Paraphrased from *SD*-I, 'Gods, Monads, and Atoms', 610.
45. BBC News, 'Scientist Finds UK Water Companies use "Magic" to Find Leaks', 21 November 2017 <http://www.bbc.co.uk/news/uk-england-oxfordshire-42070719> accessed 5 April 2020.

46. Matthew Weaver, 'UK Water Firms Admit Using Divining Rods to Find Leaks and Pipes', The Guardian, 21 November 2017 <https://www.theguardian.com/business/2017/nov/21/uk-water-firms-admit-using-divining-rods-to-find-leaks-and-pipes> accessed 5 April 2020.
47. Paraphrased from *SD*-II, 'Conclusion', 794.
48. Rupert Sheldrake, 'Is the Sun Conscious?' [video], *Renegade Tribune,* 7 July 2018 <http://www.renegadetribune.com/rupert-sheldrake-is-the-sun-conscious>accessed 6 April 2020.
49. *SD*-I, 'Summing Up – The Pith and Marrow of Occultism', 274.

Epilogue

Towards Immortality

The physical sciences are based upon, and constrained by, assumptions that have hardened into dogmas so that the scientific paradigm, or what is popularly known as the 'scientific worldview', is becoming increasingly outworn and in need of a reformation. And that reformation comes, ironically, from science itself. By its own methods and discoveries, spurred by its devotion to investigating truth, science is now transcending the materialist philosophy and pointing towards a new sense of a living world whereby the universe is no longer regarded as a machine running down towards its 'heat death', but—rather like our own Earth, Gaia—as a living, evolving organism with an inherent memory. These new paradigm shifts in the sciences shed fresh light on the whole question of spirituality and spiritual practices and also on paranormal phenomena.

The road ahead for science is steep and narrow in its unavoidable ascent towards realising that: (*a*) the origin of all life and existence is spiritual, not material; (*b*) the intrinsic nature of the universe, and all beings, is not material but mental in essence; and (*c*) consciousness in man is not generated by the brain.

The marvels of modern technology have, unfortunately, mesmerised people into believing that modern science is the only valid pathway to discovering the truth about ourselves, nature and the universe.[1] Related to that, religion (whose language is symbolism and allegory) has been banished because of our symbolically illiterate culture characterised by an excess of quantitative measures over qualitative values and by literalism and fundamentalism in religion and also the sciences.[2]

There is arguably no finer example of the primacy of matter in the scientific mainstream mind and its obsessive matter-centric worldview than the current project at the European

Organization for Nuclear Research (CERN) to build the next gigantic particle collider. Known as the Future Circular Collider (FCC), it will be a 100-kilometre ring-shaped particle accelerator buried underground near Geneva.[3] It will be many times more powerful than the current largest scientific instrument in the world—the Large Hadron Collider (LHC) at CERN contained in a 26.7-kilometre diameter tunnel.

It was the LHC that discovered the Higgs boson in 2012[4] almost half a century after its initial theoretical prediction in 1964 by the British Nobel physicist Peter Higgs FRS (1929–2024). This is an elementary particle in the Standard Model of particle physics, but it appears that many questions about the universe still remain unanswered. The Standard Model of Big Bang cosmology is inadequate to explain the remaining ninety-five per cent thought to be made up of dark energy and dark matter. The FCC project is designed to discover new particles to seek answers to the kind of questions posed above about our universe.

The total operating budget of the LHC runs to about US$1 billion per year; and the total cost of finding the Higgs boson ran to about US$13.25 billion.[5] In comparison, the cost of the FCC (not due to be completed until 2040) is expected to be US$23.8 billion.[6] A similar project is supposedly planned in China.

The reason why the current global culture is not propitious 'to wither the preachers of a mad materialism with scathing scorn'[7] is largely to do with the power-and-prestige-driven egos of scientists.

These savants of science cherish a blind belief that they wish to impose upon society that science deals only with proven facts and completely explains everything—or will do so in the near future. This is known as 'promissory materialism', a term coined by the Austrian-born British philosopher Sir Karl Popper FRS (1902–1994). And yet physicists openly acknowledge that they know only around five per cent about the total matter in the universe and so there seems to be an awfully long way to go to realise such a dream.

Science attempts to explain away phenomena completely beyond its legitimate field of discourse. It asserts that immortality of the soul and spirit is a figment of the naïve imagination and that extinction of consciousness upon death is a scientifically proven fact since mind and consciousness are emergent properties of the brain.

Modern science and technology is mastering the physical universe and building huge industries, but with a few glorious exceptions from the enlightened amongst its ranks, are completely unaware of that mysterious fount of conscious power within each living being. Without that power there would be no scientific research, no industries, no genome research, no COVID-19 vaccines, no Large Hadron Colliders, no rockets to Mars, no transhumanist attempts to attain immortality and not even such passionate zeal by atheist scientists to write books about our ultimately purposeless life in a godless universe.

All over the world men and women, worn out by the numerous, soulless cultural systems of today are crying out for the return of a lost age of beauty and enlightenment—for something practical in the highest sense of the word. A few are beginning to realise that so-called civilization in its present form is certainly at a turning point if not a vanishing point, that heartless commercialism and material efficiency are impractical and only that which offers opportunity for the expression of truth, love, and wisdom is truly worthwhile.

Not knowing where happiness truly lies, many people, especially youngsters, are flailing around, lost in a miasma of despair caused by separation from their essential natures. Hence, they become ready prey to the glitter of gain, duped in large measure by the leaders of nations preaching never-ending economic growth as the royal highway to Shangri-La. But not only this... tied into the vast economic and political systems, medical technology is feted as the 'magic bullet'—the cure all—for all mankind's health problems. Man gazes at the Medusa-like face of greed and, standing petrified, looks either to conventional science and medicine as his only salvation or seeks various forms of escape from himself.

There is no doubt that the 2020 COVID pandemic was a resounding wake-up call to mankind to reorient his sense of values away from unbridled commercialism towards a more natural lifestyle with greater reverence towards his fellow beings, nature and the planet he inhabits.

The ultimate objective of billionaire businessman Elon Musk is seeing human life on Mars.[8] Having polluted, over-populated and basically made a mess of living on our dear planet Earth, he proposes that we move on to another planet and start afresh.

The American industrialist Jeff Bezos (*b.*1964) has other ideas about how his wealth can save humanity. Blue Origin, his space company, is currently focussing on space tourism 'but he is preparing for a future in which most humans will live in floating space colonies with Earth-like features, mining and heavy industry is conducted on other planets *and Earth becomes a national park* [emphasis added]. It is the only way, he says, of ensuring rising living standards for an expanding human population without destroying Earth.[9]

By the way, who has granted Bezos divine rights to extract minerals from other planets and supposedly pollute their environments through 'heavy industry'?

The sheer ignorance and naïveté of such super-egocentric thinking by supposedly intelligent people (especially plutocrats) beggars belief. The cardinal error is to think that by changing outer circumstances the inner life will improve. It is completely the other way round. *The unequivocal teaching of sages since time immemorial is that only by a radical change in the inner structure of consciousness can we change the outer conditions of our lives.*

This is precisely why in this increasingly commercial age, mainstream science is still largely concerned with the cold classification of physical facts and knowledge, technological progress and the investigation of the temporal and illusionary parts of nature. (Technical jargon is occasionally invented to mask ignorance.) That which concerns man's soul nature and the purpose of his life is supposed to be the stuff of idle dreams, unrelated to the practicalities and economics of daily life.

Certainly, science and technology have greatly alleviated physical suffering and improved material conditions in all manner of means but have they added to the overall stock of human *happiness*?

Religion, too, has become materialistic in a different sense. Having all but lost the truly life-affirming guidance imbued esoterically in the scriptures, it remains stuck to outworn exoteric concepts of a bygone age, the beauty and dignity of faith measured by largely empty prattle and dogma from the pulpits of churches, mosques, and temples. That is the main reason why one frequently hears scientists making belittling remarks like: 'Religion is based on faith whereas science is based on facts.'

Orthodox Christianity owes the devil gratitude for 'his' existence. Having cunningly personified what is the impersonal, counteracting force in nature, the tracts of real estate and overflowing coffers of the Church are due to its exercise of power and control over unthinking people. Such people, in good faith, believed—and still do believe—that allegiance and material donations to the Church will alone afford protection from Satan corrupting their souls. In short, they imagine that they can buy salvation.

How about philosophy? Like other branches of human thought, philosophy has been made 'practical' and 'rational' (again mistaking the utilitarian for the practical and the cerebral for the rational) and its activities have become so denigrated that they have been prostituted to the material.

And what about the social science that studies the production, distribution, consumption of goods and services and transfer of wealth—economics? The vast majority of economists are concerned with the balance between capital and labour, with land and environmental concerns relegated to second place. Are economic and business models, then, based on entirely materialistic and fiscal considerations bereft of ethical values? With rare exceptions, it would seem so.[10]

The tides of evolution, however, are not at a scientist's behest as some transhumanists seem to imagine. In the future mankind will realise that the hidden purposes are none other than what occult science has taught since the dawn of humanity. And, this royal science has taught the overriding influence of greater 'parental' intelligences from planetary and solar schemes higher than our own ('gods', relatively speaking) watching over and guiding the spiritual evolution and consciousness of our humanity.

Two Paths

In relation to evolution and progress, man has a choice of two paths in life: the 'Path of Matter' or the 'Path of Spirit'. One path will always prevail and ultimately predominate. The implications of the choice are momentous.

Those who are instinctively of the nature of matter will choose the Path of Matter otherwise known as the 'Path of the Shadows'. Those who tread this path are referred to in occultism as 'Brothers of the Shadow'. Their motives are essentially self-seeking. Any extraordinary powers and faculties attained are turned to personal gain for power and control. Hence, when taken to extremes, they can descend to the Left-Hand Path of sorcery or black magic.

Multitudes of human beings are unconsciously treading the Path of the Shadows and only relatively few self-consciously follow the Right-Hand Path of self-renunciation and self-conquest, compassion, and expansion of heart, mind and consciousness towards philanthropic service. These few are known as the 'Sons of Light'.

These two classes can be distinguished between the authoritarians and the liberals—and both would seem to be increasingly staunch in their views.[11] The extreme end of the first class is characterised by a secular, aggressively anti-religious attitude and a materialistic, even mechanistic, philosophy of life. Hence, they seek the security and authority of structured environments with rules and protocols, algorithms and artificial intelligence. Unsurprisingly, their outlook is one of imposing a global uniformity (philosophically, socially, politically, etc.). They prefer to be convinced by lots of data rather than personal experience. Progress is seen primarily in terms of technological advancement, material prosperity, and economic growth.

In contrast, many of the second class are religiously or spiritually inclined. Their individuality thrives in an environment of freedom and independence and they instinctively understand the implications of diversity within unity. A distinguishing hallmark of the two classes is their attitude towards science and technology. The first tends to use the latest scientific discoveries for personal gain and power, along with dominance over others and nature, in order to prolong life by artificial and synthetic means. The second seeks to liberate society and aspires to a higher worldview than materialism. But it must be stressed that most people are somewhere around the middle of these antithetical extremes.

In this short book we have covered a vast terrain in our narrative about the primacy of consciousness and its unfolding at all levels from the spiritual to the physical and through all the kingdoms of nature. Journeying over difficult territory at times, we have shown how the whole world of science is feverishly seeking an answer to the mystery of Consciousness, spending vast sums in research, producing acres of learned papers, inflicting untold suffering upon innumerable animals ... yet all the while each of us—each single one of us without exception—is literally bearing that secret that he seeks.

If we were privileged to examine the beauty of a pearl beyond price in our hand, would we knowingly throw it to swine to trample on; or else, go looking for it far and wide in the four corners of the Earth—or would we simply look at it in the very hand that holds it?

Where, then, does that 'pearl' lie for each one of us? In all that huffing and puffing to build another gigantic machine that will reveal 'the ultimate secrets of the universe'? In the innumerable concepts woven out of our brain fabric that will soon solve the mystery of 'the origin of consciousness'? Or have we finally come to realize the counsel of the illimitable sages of both the East and the West: our body is the Path, and Consciousness is the Traveller. The age-old enigma of where the Pearl of Wisdom lies is Within Oneself—*Thou art THYSELF the object of thy search.*[12]

NOTES

1. See Kingsley L. Dennis, *The Sacred Revival.*
2. See Avinash Chandra, *The Scientist and the Saint: The limits of science and the testimony of sages* (Cambridge, UK: ArcheType Books, 2018).
3. 'Future Circular Collider', CERN <https://home.cern/science/accelerators/future-circular- collider> accessed 6 April 2020.
4. 'The Higgs Boson', CERN <https://home.cern/science/physics/higgs-boson> accessed 6 April 2020.
5. Alex Knapp, 'How Much Does It Cost To Find A Higgs Boson?' *Forbes* (5 July 2012) <https://www.forbes.com/sites/alexknapp/2012/07/05/how-much-does-it-cost-to-find-a higgs-boson/#188ddbcb3948> accessed 6 April 2020.
6. Davide Castelvecchi, 'Next-generation LHC: CERN lays out plans for €21-billion supercollider. The proposed facility would be the most powerful collider ever built',

Nature (15 January 2019) <https://www.nature.com/articles/d41586-019-00173-2> accessed 6 April 2020.

7. *NPB-8*, *Reflections*, 'The Literary Work', ¶184.
8. Staff writer, 'Elon Musk Unveils SpaceX Rocket Designed to Get to Mars and Back', 30 September 2019, *news.com.au*.
9. 'Jeff Bezos: The Amazon billionaire's galactic ambitions', *The Week*, 13 February 2021, 22.
10. Brian Hodgkinson, *How our Economy Really Works: A radical reappraisal* (London: Shepheard- Walwyn, 2019). Refer also to the books listed under <https://shepheard-walwyn.co.uk/ product- category/ethical-economics> accessed 24 June 2020.
11. There is an insightful discussion on this in, 'Polarised: Exploring divisions', RSA Journal, 2 (2018), 6.
12. *VS*, Fragment I, ¶99.

About the Author

Edi Bilimoria DPhil, FIMechE, FEI, FRSA

Esotericist and prominent engineer, Dr Edi Bilimoria is a modern-day polymath uniquely combining a profound understanding of philosophy with cutting-edge science and the arts.

Born in India and educated at the universities of London, Sussex and Oxford, he has had a distinguished career as a consultant engineer to the petrochemical, oil and gas, transport, and construction industries. He was project manager and head of design for major innovative projects such as the Channel Tunnel, London Underground systems, offshore installations, and safety and environmental management for several Royal Navy projects, including the Queen Elizabeth Aircraft Carrier.

A student of the perennial philosophy for over half a century, Edi has lectured extensively in the UK, California, the Netherlands, India, and Australia. He has organized and chaired conferences in order to encourage the cross-fertilization of ideas in the fields of science, religion and practical philosophy.

Edi has published many informative articles and papers in the disciplines of science, engineering, and esoteric philosophy. His book *The Snake and the Rope* was awarded the Book Prize by the Scientific and Medical Network (SMN). In 2023, the full four volume version of this book was awarded the SMN's Grand Prize. Applauded by many, it is considered to be the most penetrating and all-embracing work on consciousness written in the modern era.

A highly-experienced glider pilot, Edi is also a choral singer and a dedicated pianist of concert standard.

Previous Books by Dr Edi Bilimoria

Unfolding Consciousness: Exploring the Living Universe and Intelligent Powers in Nature and Humans
Publisher: Shepheard-Walwyn
ISBN: 978-0-856835-353

The Snake and the Rope: Problems in Western Science Resolved by Occult Science
Publisher: Quest Books/Theosophical Publishing House
ISBN-10: 817-059-4847
ISBN-13: 978-817-059-4840

A Personal Note to Readers

There are splendid books on science, religion, occult science, philosophy, and art; but few that have brought these diverse streams under a single overarching theme comprising a single body of integral wisdom whilst pointing to their source. Allied to this, a constant theme of this book is the universality of the Mystery teachings of all cultures and religions. That does not mean they are all the same, but they are in harmony and confluence one with the other and, also, with enlightened science. This is not science in the narrow sense, the knowledge of phenomena and the objective world appreciable to our physical senses, but science in its larger and truer sense. This science discloses the radiant truth because it is free from all man-made accretions and superstitions.

I have avoided the all-too-common tendency amongst writers of stating 'the truth' on any particular subject as a *fait accompli*. I have sought to argue *towards* the truth but to leave readers entirely free to come to their own conclusions.

In the wake of over half a century of intense involvement and participation in spiritual societies and musical circles, contemporaneous with a working life in diverse roles across numerous industries, I ask the reader's indulgence in permitting me to pass on four observations hard-won through experience.

First, the need to transcend—never jettison—intellect in order to arrive at even an approximate semblance of truth. We live in a world where reality is ascribed to the surface appearance of objects. However, the noumenal essences have more reality than their phenomenal, objective counterparts. We cannot approach truth merely through science and the intellect.

I am unconditionally and utterly convinced as to the ubiquity and primacy of consciousness. Moreover, all my research leads to the same conclusion, namely—that consciousness is transmitted through the human being and filtered by the brain, its instrument of expression. This is, of course, contrary to the current virtually unanimous mainstream neuroscience diktat that thought is produced by the brain and consciousness is nothing more than a distracting epiphenomenon—'ghost in the machine'—explained away by physical processes.

In my view, whereas there is never any question of artificial intelligence overtaking human intelligence, I regard the blurring of the two by virtue of the increasingly rampant glorification of the former, along with the 'machinisation' of human intelligence, as one of the most pernicious trends in consciousness research by contemporary mainstream science. The problem is compounded by the promise of a digital afterlife and gene reprogramming in order to live forever. With notable exceptions, sadly, such research is still driven by materialistic concepts underpinned by unsupportable presuppositions, at the cost of reality and truth.

On a brighter note, there is a burgeoning movement amongst increasing numbers of enlightened scientists and philosophers investigating such research evidence that has been ignored or dismissed because it is philosophically incompatible with materialism.[b] In this we see reflected a heartening attempt to build bridges between spirit, mind and matter. This movement argues for global dialogue about a science of consciousness without restricting consciousness solely to a sub-discipline of mainstream neuroscience. It further proposes that consciousness can give us direct access to the deeper structures of reality and therefore stresses the importance of coupling the normal theoretical third person perspective approach to consciousness studies with the experiential, first-person perspective.

---

b For example: The Scientific and Medical Network, https://scientificandmedical.net/ • Institute of Noetic Sciences (IONS), https://noetic.org/ • Society for Psychical Research, https://www.spr.ac.uk/

Second, the indispensable need always to maintain a universal outlook, which demands a plurality of approaches and universality of outlook which draws on diverse wisdom streams, whilst always endeavouring to discover their unifying thread and seeking out their Source. Sectarianism and its twin, dogma, are the greatest curses that have blighted not only religion, the obvious example, but, albeit more covertly, science, philosophy, and esotericism no less. I maintain categorically that there is no one book, teacher or teaching, of any epoch, that can be upheld as an absolute gold standard of truth.

That sectarian attitudes, like poisonous mushrooms, pervade spiritual societies which are always founded on the basis of *truth wherever it leads*, is quite deplorable. To maintain such a stance (promulgated in the guise of upholding the 'purity' of the original teachings) is nothing short of fundamentalism and indicative of weak minds—the equivalent of religious orthodoxy. Attempting to establish an ultimate 'Church of Truth' based on one authority, whose say-so is propped up by its self-appointed popes or archbishops, would be an affront to truth. Furthermore, truth is best served by both intellect and intuition—the way of science in hand with the path of mysticism.

Third, it is a huge mistake to believe (as I credulously did for some decades) that members of spiritual organisations are by definition a 'cut above' the common folk in terms of their generosity, sincerity, and altruism. Human nature is always the same and I have personally experienced genuinely fraternal behaviour as much from those in the wide world of industry, and from friends and colleagues who would not have any clue about esoteric science, as from those few in spiritual organisations.

Whereas politics, corruption, and skulduggery are nothing new to commercial organisations, they are no less extant and, in fact, more insidious in spiritual or esoteric societies than the secular world. This is because such societies were set up for the express purpose of promulgating truth and therefore have a sacred duty in this respect. If departed from, for self-seeking motives, this constitutes a greater fall and dereliction of duty than would be the case in ordinary secular life. The amount of backbiting and infighting in such societies—amongst those who profess such things as brotherhood and universal love—is an adulteration and falling away from the nobility of the ideals and spiritual vision of their founders. As Mahatma Gandhi counselled, the best way to change society (let alone the world) is first to start with oneself—and that is rarely achieved merely by piously quoting long passages about 'service to humanity' from sacred texts or revered books at public lectures.

Finally, there is the need to listen to all but to maintain one's own counsel. 'Art thou able to walk alone?' was a testing question once put to the contemporary sage and philosopher Paul Brunton in his younger days.[c] Those who are not sure about how to answer might care to take to heart this extract from another adept, the Tibetan Djwal Khul: 'Be prepared for loneliness. It is the law. This must be endured and passed... [Additionally] have patience. Endurance is one of the characteristics [of the aspirant].'[d] Whatever societies we may or may not belong to, we can listen to all but must ultimately walk our own path—alone.

Edi Bilimoria, Godalming, Surrey, UK, 23 April 2024

c Paul Brunton, *A Search in Secret Egypt* (India: B. Publications, 1935), 71.

d Alice Bailey, *Initiation, Human and Solar* (UK: Lucis Trust), 76.

Glossary and Definitions

P = Pali S = Sanskrit

Akasha (S)

Literally, 'shining', or 'luminous'; hence, the primordial, supersensuous spiritual essence which pervades all space. It is also known as Æther but it is not the ether of science which is one of its lower elements. For this reason, akasha constitutes the 'tablet of memory', also referred to as the 'Akashic Record', or 'Akashic Chronicle', the enduring record of all that happens, and has ever happened, in the whole of the universe. (*See also* Astral Light.)

Antahkarana (S)

Literally, the 'internal instrument', the intermediate instrument functioning as bridge, or medium of communication, between the Lower and Higher Manas. In this sense, therefore, a vehicle of consciousness.

Astral body

See Linga-sharira.

Astral Light

The lower levels of Akasha. The inner seat of all memory, a record of all that has happened. It is in the Astral Light that the memory of man and nature is 'stored' indelibly and virtually forever—and not, in the case of man, in his fleshly brain which acts as a retrieving mechanism. (*See also* Akasha.)

Atma (S)

The Divine Self, 'Pure Consciousness', otherwise known as Universal Spirit or Supreme Soul. The divine (highest) aspect in man's constitution. It is that essential and radical faculty or power which gives every entity its knowledge and sentient consciousness of selfhood.

Avitchi (S)

Literally, 'waveless', referring to a state of the greatest isolation and stagnation attained after physical death as the result of a life of utter evil. Also a generalised term for places of evil realization (but not of 'punishment' in the orthodox Christian sense) where the will for evil and unsatiated evil desires find their chance for expansion—before final extinction of the entities.

Buddhi (S)

As the vehicle of Atma it is the spiritual soul providing the faculty of discrimination and intuition that awakens man to direct understanding. Buddhi in conjunction with Atma constitutes the Human Monad.

Causal body

The union of Buddhi with Higher Manas. It is also referred to as the Karmic Body.

Cosmos

The term 'Cosmos' written with an upper case *C* refers to our solar system, unmanifest and manifest, subjective and objective, to emphasize its ordered and harmonious characteristic.

The term 'cosmos' written with a lower case *c* refers to the objective aspect of Cosmos, that is, the manifest, objective aspect of our solar system comprising the sun and the celestial objects that orbit it, either directly (like the planets) or indirectly (like the planetary moons).

Devachan (S)

Literally, the 'dwelling of the gods', being a spiritual state intermediate between two Earth- lives into which the human monad enters and there rests in repose and bliss; and the fulfilling of all *unfulfilled* spiritual yearnings of the past incarnation before the subsequent urge to reincarnate. It is not identical with the Heaven of orthodox Christianity conceived of as a permanent state.

Kama (S)

The principle or state of desire, expressing itself, particularly in man as the motivating principle and impelling force behind emotion and passion.

Kama-loka (S)

Comparable to Hades in Greek mythology and purgatory in Christian theology, an intermediate state of consciousness after death in which the kama-rupa of the departed must undergo purification so as to achieve the holiness necessary to enjoy the consciousness of devachan. (*See also* Kama-rupa, Devachan.)

Kama-rupa (S)

The subjective form created by means of the thoughts and desires of a person during life projected as a form into the astral world after the death of the physical body.

Karma (S)

Physically, action: metaphysically, the universal law of retribution, the law of cause and effect, or ethical causation. Karma is neither punishment nor reward *per se*, but becomes either depending on the cause that generated the logical effects. There is the karma of merit and the karma of demerit.

Kosmos

The term 'Kosmos' written with an upper case *K* refers to the numberless, infinite Universes, the innumerable Solar systems of the infinite Cosmos, unmanifest and manifest, subjective and objective.

The term 'kosmos' written with a lower case *k* refers to the objective aspect of Kosmos. The term 'multiverse' in scientific cosmology approximates to kosmos.

Linga-sharira (S)

The model body, or astral body, as the causal form, or template of the Physical body.

Manas (S)

Mind as a principle. The mental faculty which makes of man an intelligent and moral being, and distinguishes him from the animal. It is the pivot point bridging the Individuality with the Personality. A dual function in man, Manas can either 'rise' to ally with Buddhi (the Higher Mind) or 'descends' to identify with Kama (the Lower mind).

Maya

Literally, subject to change, hence appearance, or illusion.

Mayavi-rupa

The 'illusive form', 'illusory body', 'illusion body', 'dream body', 'thought body', all such terms indicative of a higher astral-mental form. Similar terms in German and French are *doppelgänger* **and** *perispirit*, respectively. Note that the māyāvi-rūpa pertains to the living person.

Monad

Literally, an ultimate unit of being, eternal life-centres, consciousness-centres; the Unit Spirit in or overshadowing everything. In man, the conjunction of Atma and Buddhi. (*See* Atma, Buddhi)

Prana (S)

The breath of life, the vital principle, or life-force, that pervades and animates the physical body.

Rupa (S)

Literally, 'form', or 'body' (as opposed to arūpā, 'formless', or bodiless').

Shamballa (S)

A place-name of highly mystical significance where, according to the most ancient traditions, lived those great intelligences who are the real guides of our planetary life as a whole, keeping watch over the evolution of the various kingdoms of our planetary nature, seen and unseen—not just mankind.

Sthula-sharira (S)

The vehicle of the six non-physical principles; that which gives expression in the physical world of all man's faculties during life.

Tanha (P)

The 'thirst' for material life; the desire to live and to cling to life on this Earth (being the main cause of rebirth or reincarnation). (*See also* Trishna.)

Trishna (P)

Literally, 'thirst' and 'longing' for what a man formerly knew and what he wills and desires to know again—things familiar to it and akin to it from past experiences— which draws the man back again to incarnation on the Earth plane. (*See also* Tanha.)

Further Reading

Besant, A., *Seven Great Religions*, Adyar, Madras: Theosophical Publishing House, 1972.

Brunton, P., *The Quest of the Overself*, London: Rider & Co., 1937.

Cranston, S., *Reincarnation: The Phoenix Fire Mystery – An East-West Dialogue on Death & Rebirth from the Worlds of Religion, Science, Psychology, Philosophy*, Theosophical University Press, 1994.

Elsaesser, E., *Spontaneous Contacts with the Deceased*, John Hunt Publishing – IFF Books, 2023.

Farthing, G. A., *Deity, Cosmos and Man: An Outline of Esoteric Science*, Point Loma Publications, 1993.

Fenwick, P. and Fenwick, E., *Past Lives: An Investigation into Reincarnation Memories*, Headline Book Publishing, 1999.

László, E., *Science and the Akashic Field: An Integral Theory of Everything*, 2004; 2nd edn, Rochester, Vermont: Inner Traditions, 2007.

Leadbeater, C. W., *Man Visible and Invisible*, Zinc Read, 2023.

Lorimer, D., *Thinking Beyond the Brain*, Floris Books, 2001.

McGilchrist, I., *The Master and His Emissary: The Divided Brain and the Making of the Western World*, 2010; rev. and enl. 2nd edn, New Haven and London: Yale University Press, 2019.

Phillips, S., *ANIMA: Evidence of a Yogic Siddhi – Remote Viewing of Particles*, Theosophical Publishing House, 1996.

Shackleton, E., *South: The Endurance Expedition*, London: Penguin Classics, 1914.

Sheldrake, R., *A New Science of Life: The Hypothesis of Formative Causation*, Blond and Briggs, 1981.

Sheldrake, R., *The Presence of the Past*, Fontana, 1989.

Taimni, I. K., *Science and Occultism*, Theosophical Books Ltd, 1974.

Warcup, A., 'An Inquiry into the Nature of Mind', the Blavatsky Lecture, 1981, London: The Theosophical Society.

Warcup, A., *Cyclic Evolution: A Theosophical View,* Theosophical Books Ltd, 1986.

Wyatt, T., *Cycles of Eternity: An Overview of the Ageless Wisdom,* Firewheel Books, 2016.

Index

ACADEMIC AND SPECIALIST

Iff Books publishes non-fiction. It aims to work with authors and titles that augment our understanding of the human condition, society and civilisation, and the world or universe in which we live. If you have enjoyed this book, why not tell other readers by posting a review on your preferred book site.

Recent bestsellers from Iff Books are:

Why Materialism Is Baloney

How true skeptics know there is no death and fathom answers to life, the universe, and everything

Bernardo Kastrup

A hard-nosed, logical, and skeptic non-materialist metaphysics, according to which the body is in mind, not mind in the body.

Paperback: 978-1-78279-362-5 ebook: 978-1-78279-361-8

The Fall

Steve Taylor

The Fall discusses human achievement versus the issues of war, patriarchy and social inequality.

Paperback: 978-1-78535-804-3 ebook: 978-1-78535-805-0

Brief Peeks Beyond

Critical essays on metaphysics, neuroscience, free will, skepticism and culture

Bernardo Kastrup

An incisive, original, compelling alternative to current mainstream cultural views and assumptions.

Paperback: 978-1-78535-018-4 ebook: 978-1-78535-019-1

Framespotting

Changing how you look at things changes how you see them

Laurence & Alison Matthews

A punchy, upbeat guide to framespotting. Spot deceptions and hidden assumptions; swap growth for growing up. See and be free.

Paperback: 978-1-78279-689-3 ebook: 978-1-78279-822-4

Is There an Afterlife?

David Fontana

Is there an Afterlife? If so what is it like? How do Western ideas of the afterlife compare with Eastern? David Fontana presents the historical and contemporary evidence for survival of physical death.

Paperback: 978-1-90381-690-5

Nothing Matters

a book about nothing

Ronald Green

Thinking about Nothing opens the world to everything by illuminating new angles to old problems and stimulating new ways of thinking.

Paperback: 978-1-84694-707-0 ebook: 978-1-78099-016-3

Panpsychism

The Philosophy of the Sensuous Cosmos

Peter Ells

Are free will and mind chimeras? This book, anti-materialistic but respecting science, answers: No! Mind is foundational to all existence.

Paperback: 978-1-84694-505-2 ebook: 978-1-78099-018-7

Punk Science
Inside the Mind of God
Manjir Samanta-Laughton
Many have experienced unexplainable phenomena; God, psychic abilities, extraordinary healing and angelic encounters. Can cutting-edge science actually explain phenomena previously thought of as 'paranormal'?
Paperback: 978-1-90504-793-2

The Vagabond Spirit of Poetry
Edward Clarke
Spend time with the wisest poets of the modern age and of the past, and let Edward Clarke remind you of the importance of poetry in our industrialized world.
Paperback: 978-1-78279-370-0 ebook: 978-1-78279-369-4

Readers of ebooks can buy or view any of these bestsellers by clicking on the live link in the title. Most titles are published in paperback and as an ebook. Paperbacks are available in traditional bookshops. Both print and ebook formats are available online.
Find more titles and sign up to our readers' newsletter at
www.collectiveinkbooks.com/non-fiction
Follow us on Facebook at
www.facebook.com/CINonFiction